Landscape Integration Design

一体化景观

（澳）乔纳森·哈弗斯　主编
凤凰空间　　译

江苏科学技术出版社

CONTENTS

NEO-CLASSICISM

MODERN AND NATURAL STYLE

ORIENTAL ZEN

FUSION STYLE

NEO-CLASSICISM

Neo-classicism is a diversified way of thinking. It retains both traditional historical elements and cultural factors. By simplifying miscellaneous carving and utilizing modern materials, neo-classicism abandons over complex patterns and decorations, showcasing a new style of classicality and simplicity.

Dunbier Garden

Design Company: Scott Brown Landscape Design
Location: Mornington Peninsula, Victoria, Australia
Site Area: 4,046.9 m^2
Photographer: Patrick Redmond

Sea View Pool-side Elegance

Terrain: Beach Slope

From a commanding position, this dramatic residence on the Mornington Peninsula overlooks Port Phillip Bay, with distant views of the City, as well as the heads. Most of the outdoor space lies in front of the home, with the gently rolling lawns and some garden forming a "door mat" to this view.

Russell and Kay wanted a swimming pool and Alfresco dining room to complete the picture. The requirements for the dining room were particularly critical as the existing residence has its kitchen upstairs. The visual style was left up to me...it went without saying that the pool, Alfresco dining room and kitchen, and the surrounding precinct needed to blend with the home and its existing setting, whilst preserving the amazing views. As Russell and Kay both know well, my aim in this case was to create a space where the line between the old and the new would be largely invisible...as if the entire outdoor space including the pool and Alfresco room was created simultaneously with the house.

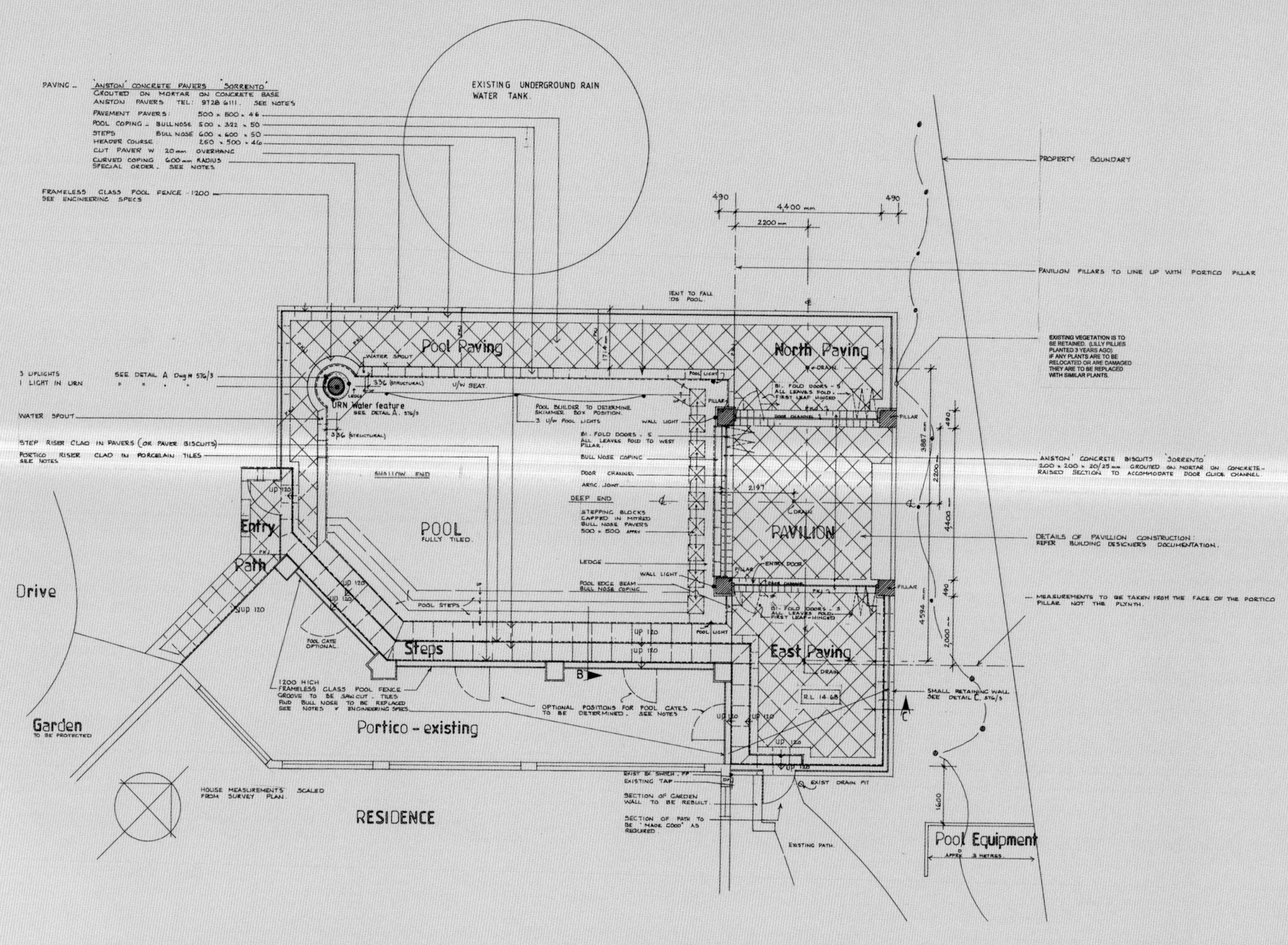

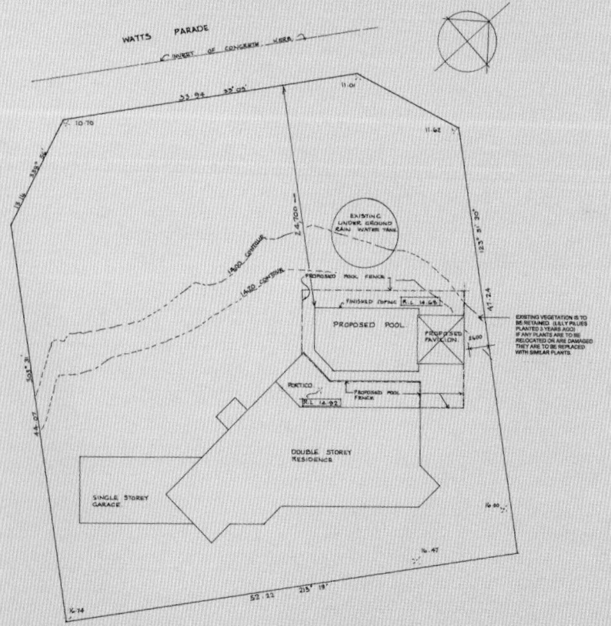

Initial considerations had to deal with the prevailing weather, particularly south-west winds and sea breezes as well as the passage of the sun. The site's magnificent vistas also meant some climatic exposure and these issues had to be managed to enable the finished spaces to be enjoyed and used without the family being held to ransom by Melbourne's fickle weather.

The angles of the home have been used to drive the design. By aligning the pool's axis with that of the main spine of the home, both the home and pool look as though they are related, as if they are part of the same picture. This underscores the physical and visual relationship between the pool and the home.

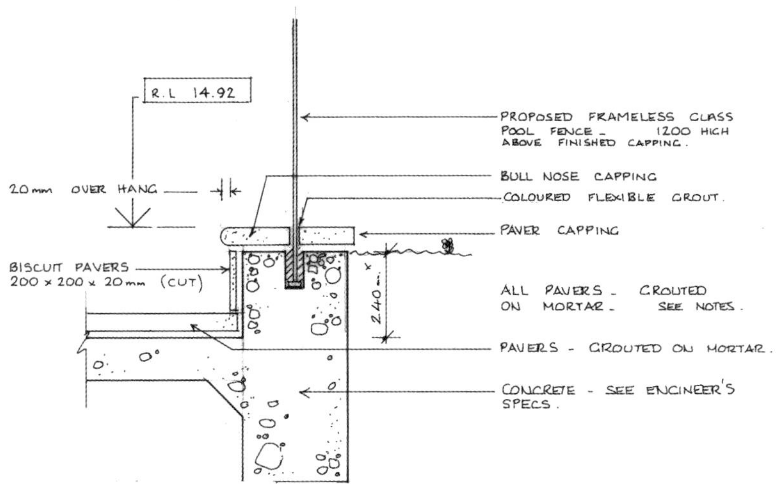
R.L 14.92
PROPOSED FRAMELESS GLASS POOL FENCE - 1200 HIGH ABOVE FINISHED CAPPING.
20mm OVER HANG
BULL NOSE CAPPING
COLOURED FLEXIBLE GROUT.
PAVER CAPPING
BISCUIT PAVERS 200 x 200 x 20mm (CUT)
240mm
ALL PAVERS - GROUTED ON MORTAR - SEE NOTES.
PAVERS - GROUTED ON MORTAR.
CONCRETE - SEE ENGINEER'S SPECS.

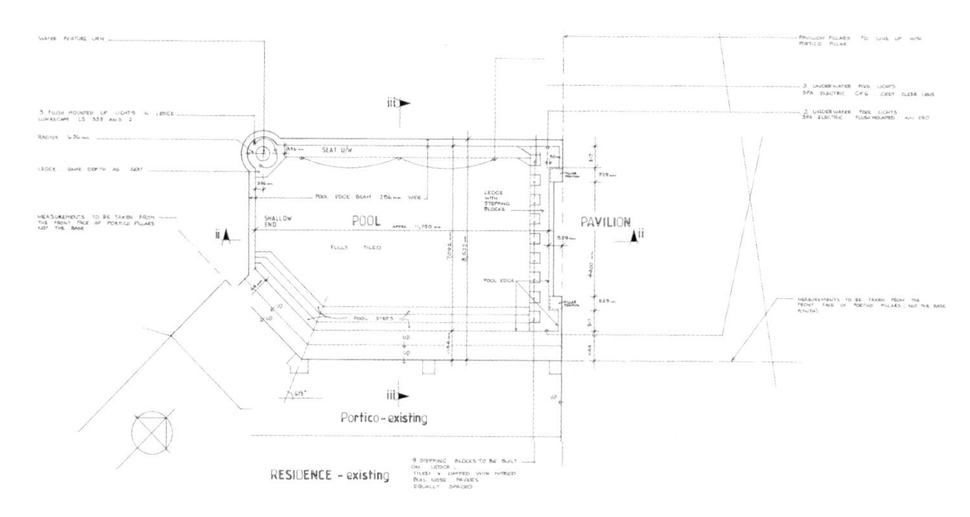

POOL
PAVILION
SHALLOW END
FULLY TILED
Portico-existing
RESIDENCE - existing

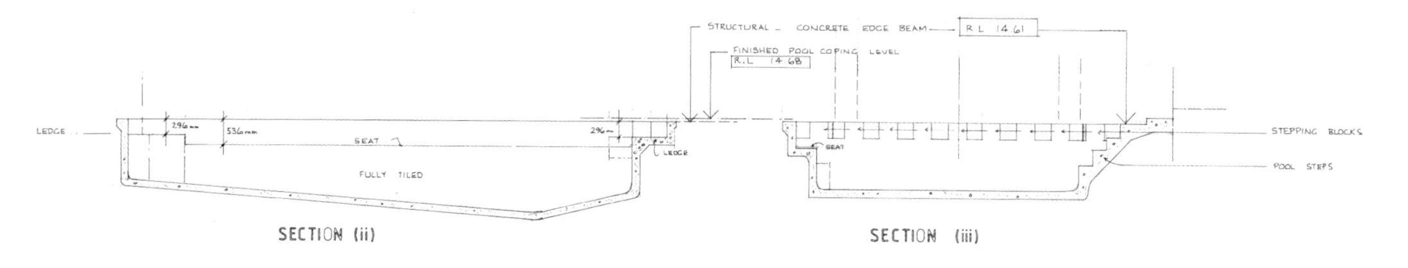

STRUCTURAL - CONCRETE EDGE BEAM
R.L 14.61
FINISHED POOL COPING LEVEL
R.L 14.68
LEDGE
SEAT
FULLY TILED
STEPPING BLOCKS
POOL STEPS
SECTION (ii)
SECTION (iii)

The home had existing patio verandas, with a small step down to the lawns. In this case, we have set the pool level two more steps lower, and ensured the pool position is close enough to the home to create a sense of them being "one", within much larger gardens and lawn areas. This also brings other advantages. The pool's width and surrounding paving is set lower, and against the home, so as not to extend out and reduce or obstruct the important vistas.

Another advantage of this "step down" is that we have been able to position the Alfresco building in close proximity to the home without having to physically join the two, even though we have replicated the architecture of the home in the Alfresco building itself. When constructing a new building very close to an existing one, the two buildings can tend to look "plonked" and unrelated, unless they are physically joined, especially when they are very similar in architecture. The change in paving level has allowed the two buildings to be independent of each other without looking "too close", because the roof levels are subtly different.

With the alignment and position of the pool conceptually decided, there were a number of technical issues to deal with. Digging such a large hole so close to a two story building would normally necessitate structurally underpinning the home so it would not collapse into the pool cavity. By designing terraced steps along the entire house side of the pool, we managed to avoid any structural underpinning requirements. Aesthetically, the steps give a great link to the home. The pool can be accessed from anywhere along the home side, and these underwater steps link to the two above-water steps at the edge of the veranda.

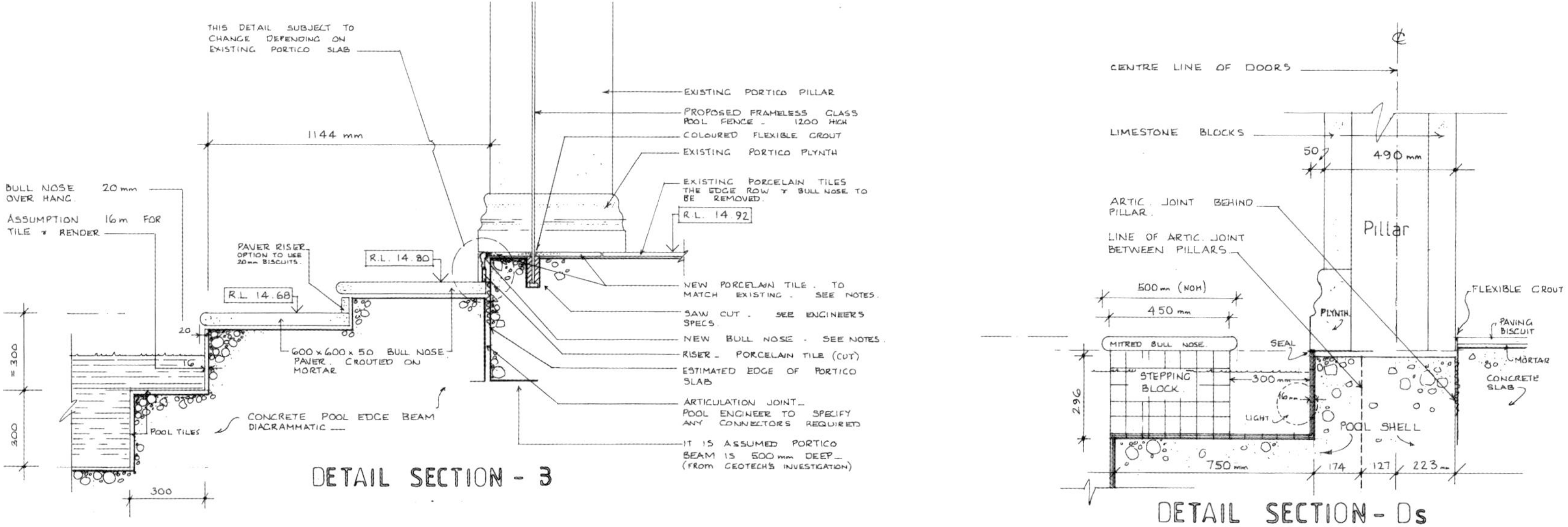

THIS DETAIL SUBJECT TO CHANGE DEPENDING ON EXISTING PORTICO SLAB
EXISTING PORTICO PILLAR
PROPOSED FRAMELESS GLASS POOL FENCE _ 1200 HIGH
COLOURED FLEXIBLE GROUT
EXISTING PORTICO PLYNTH
1144 mm
BULL NOSE 20 mm OVER HANG
ASSUMPTION 16m FOR TILE + RENDER
EXISTING PORCELAIN TILES THE EDGE ROW + BULL NOSE TO BE REMOVED
R.L. 14.92
PAVER RISER OPTION TO USE 20mm BISCUITS
R.L. 14.80
R.L. 14.68
NEW PORCELAIN TILE. TO MATCH EXISTING. SEE NOTES.
SAW CUT. SEE ENGINEERS SPECS
NEW BULL NOSE - SEE NOTES.
600 x 600 x 50 BULL NOSE PAVER. GROUTED ON MORTAR
RISER _ PORCELAIN TILE (CUT)
ESTIMATED EDGE OF PORTICO SLAB
ARTICULATION JOINT_ POOL ENGINEER TO SPECIFY ANY CONNECTORS REQUIRED
CONCRETE POOL EDGE BEAM DIAGRAMMATIC
POOL TILES
IT IS ASSUMED PORTICO BEAM IS 500 mm DEEP_ (FROM GEOTECH'S INVESTIGATION)
20
16
= 300
300
300
DETAIL SECTION - 3
CENTRE LINE OF DOORS
LIMESTONE BLOCKS
50
490 mm
ARTIC. JOINT BEHIND PILLAR.
Pillar
LINE OF ARTIC. JOINT BETWEEN PILLARS
500mm (NOM)
450 mm
PLYNTH
FLEXIBLE GROUT
PAVING BISCUIT
MITRED BULL NOSE
SEAL
MORTAR
STEPPING BLOCK
300mm
CONCRETE SLAB
16mm
LIGHT
POOL SHELL
296
750mm
174
127
223mm
DETAIL SECTION - Ds

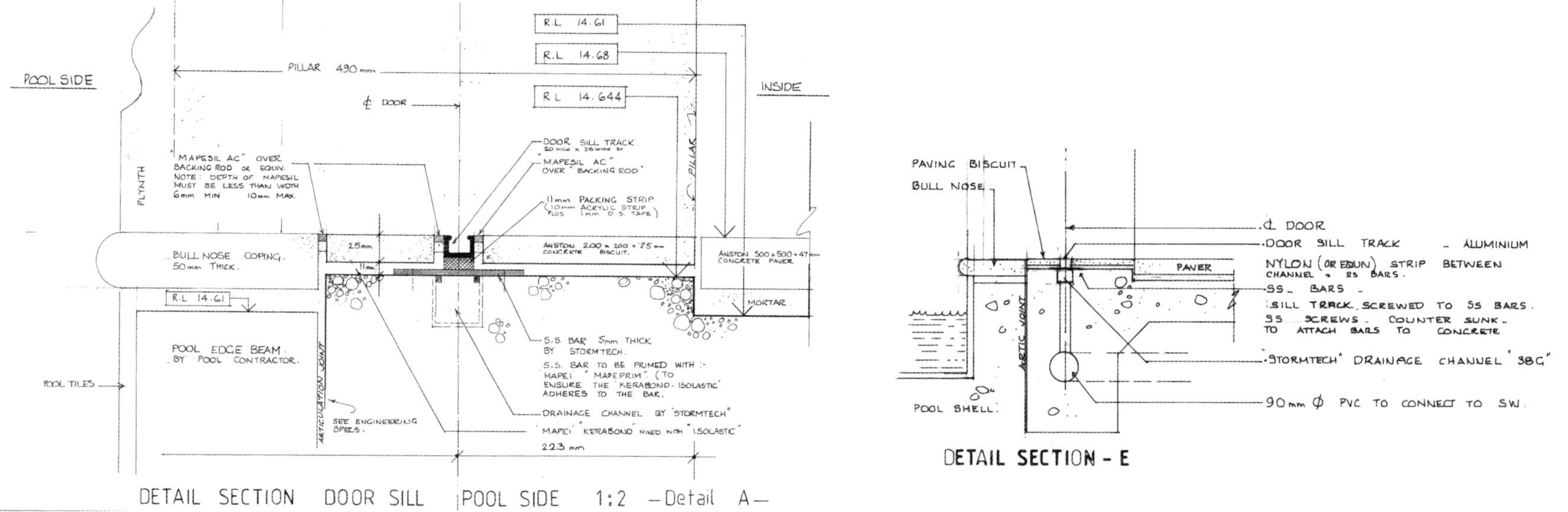

POOL SIDE
PILLAR 490 mm
INSIDE
R.L 14.61
R.L 14.68
R.L 14.644
℄ DOOR
PLYNTH
PILLAR
"MAPESIL AC" OVER BACKING ROD OR EQUIV. NOTE: DEPTH OF MAPESIL MUST BE LESS THAN WIDTH 6mm MIN 10mm MAX.
DOOR SILL TRACK.
MAPESIL AC" OVER "BACKING ROD"
11mm PACKING STRIP (10mm ACRYLIC STRIP PLUS 1mm D.S. TAPE)
25mm
11mm
BULL NOSE COPING. 50mm THICK.
ANSTON 200 x 200 x 25 mm CONCRETE BISCUIT.
ANSTON 500 x 500 x 47mm CONCRETE PAVER
MORTAR
R.L 14.61
POOL EDGE BEAM BY POOL CONTRACTOR.
POOL TILES
ARTICULATION JOINT
SEE ENGINEERING SPECS.
S.S. BAR 5mm THICK BY STORMTECH.
S.S. BAR TO BE PRIMED WITH :- MAPEI "MAPEPRIM". (TO ENSURE THE 'KERABOND- ISOLASTIC' ADHERES TO THE BAR.
DRAINAGE CHANNEL BY "STORMTECH"
MAPEI "KERABOND" MIXED WITH "ISOLASTIC"
223 mm
DETAIL SECTION DOOR SILL POOL SIDE 1:2 —Detail A—
PAVING BISCUIT
BULL NOSE
PAVER
℄ DOOR
DOOR SILL TRACK - ALUMINIUM
NYLON (OR EQUIV) STRIP BETWEEN CHANNEL & SS BARS.
SS BARS
SILL TRACK SCREWED TO SS BARS.
SS SCREWS. COUNTER SUNK. TO ATTACH BARS TO CONCRETE
ARTIC JOINT
"STORMTECH" DRAINAGE CHANNEL "38G"
90mm Ø PVC TO CONNECT TO SW.
POOL SHELL.
DETAIL SECTION - E

The Alfresco building is constructed from the same material as the home (Mount Gambier Limestone), and designed to look very similar in style. The Alfresco building is comprised of side-stacker, collapsible cafe doors on three sides, which allow for a huge number of variations and options for use...all enclosed when the winds are strong, or on a cold evening, or semi-open on a nice day when the sea breeze is still a little "fresh", or fully open on a beautiful warm summer's day (and all these options preserve the wonderful views). The position of the Alfresco building at the end of the pool provides a commanding view back along the pool and across the bay, with the gardens framing this view and even enhancing it. This positioning also means that the building does not block the views from the home. In fact, the views are even better now, as one can enjoy the views from inside the home, and feel welcomed out to the entertainment precinct, all with the magnificent bay and gardens as a backdrop.

The kitchen is set against the one solid wall of the Alfresco building, and has all the necessary elements to avoid going back upstairs to the kitchen inside the house. A full sized double fridge houses sufficient food and drinks for a small army, while the stainless steel barbeque is easy to use and is plumbed in to the mains gas supply, and the sink has hot and cold running water. The flue minimises oil build- up on the ceiling when opening the hood to "turn" the food. The client's usage of this room has grown so much that an automatic coffee machine has been added!

Climate control in the Alfresco building is provided by the heavily tinted glass of the doors (because the building has no eaves, the sun will hit the glass all afternoon), internal sun-block blinds (internal, so as not to be damaged by strong winds), and a large central ceiling fan.

The paving around the pool is a sand-blasted concrete ('Anston' – "Sorrento") which I have chosen to closely match the existing Spanish porcelain on the veranda in colour, but provide much more slip-resistance around the pool (think small grand children running around the pool). I have chosen to enclose the lower edges of the pool paving in a peninsular of garden, rather than leave it all open as lawn. By using a low growing, prostrate form of Juniper, I have given the pool area a sense of enclosure, as if it is part of the garden rather than set in the lawn, without obscuring the view.

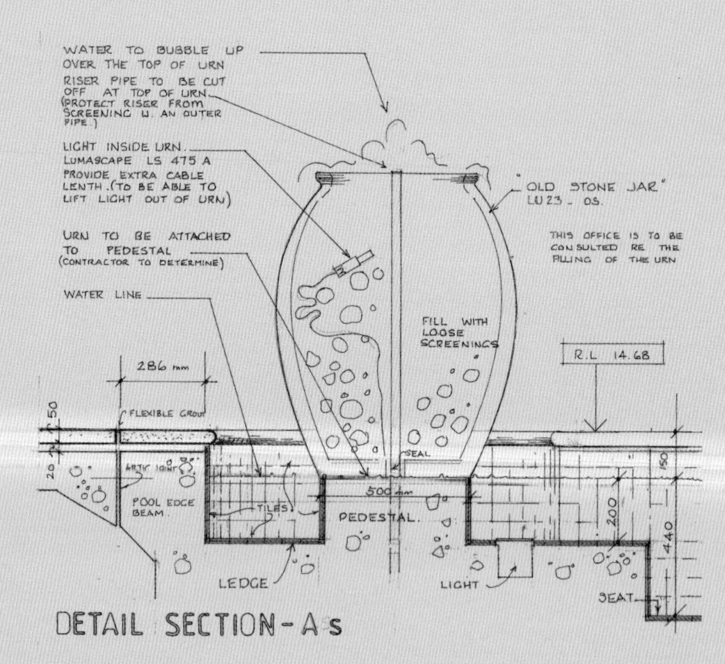

By positioning Grecian urns set in to this garden peninsular, the architectural strength of the home and its entertaining precinct has been extended into the garden.

The entire site is transformed at night with the outdoor lighting designed to enhance the drama of the home and its outdoor space.

The entire precinct is designed to sit comfortably within the existing site, and relates well with both the home and the garden setting. First-time visitors to the home have great difficulty in identifying where the original and the new stop and begin...and I take that as an indication of "mission accomplished".

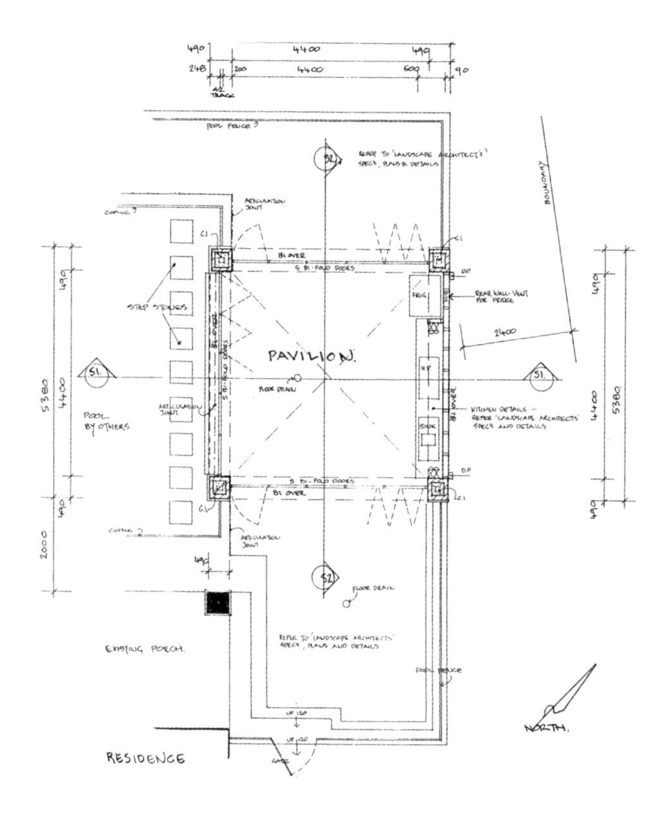
4400
PAVILION.
5380
4400
POOL BY OTHERS
EXISTING PORCH
NORTH
RESIDENCE

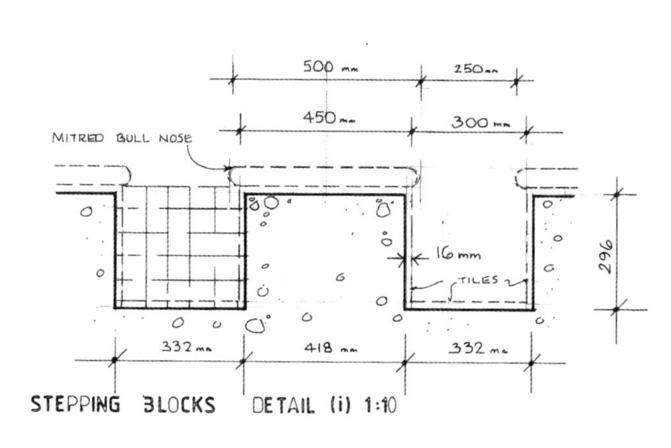
500 mm
250 mm
450 mm
300 mm
MITRED BULL NOSE
16 mm
TILES
296
332 mm
418 mm
332 mm
STEPPING BLOCKS DETAIL (i) 1:10

Private Residence 2

Design Company: Paul Sangha Landscape Architecture
Location: Vancouver, BC, Canada
Site Area: 7,082 m^2

A Neoclassical Charming Home

Terrain: Mountainous Region

This Neoclassical home, sited next to the stunning natural backdrop of Pacific Spirit Park, is a blend of unique garden rooms and experiences that complement the clients' love of nature and art. The garden brings together neoclassical principles of balance and symmetry through integration of parterre gardens, allees, and garden pavilions. The axial extensions from the house create structure for the garden, while the fluid lines act as a contrast, drawing visitors into adjoining garden rooms encouraging exploration. The transition between the formality of the garden and informality of the woodland park creates a unique juxtaposition of the urban against the natural landscape. Every path taken is rewarded with unique experiences of wild nature, water gardens, unique ornamented plantings, and lawn expanses.

The great lawn is also accessible via the main axis from the house. The grand stairs provide an elegant walkway, spilling gracefully out onto the lawn, which is lined by a series of pavilions, providing shaded areas for relaxing and dining. A hedge was placed as a horizontal foil to the park so that both the park and garden could co-exist harmoniously, all while heightening the experience of each when you moved past it. A path via the Pine Allee winds through the sea of ferns and mosses of Pacific Spirit Park to allow you to be completely immersed in the natural beauty of the park.

Each space flows seamlessly together to marry both the beauty of the surrounding park, and this elegant neoclassical home nestled amongst its sweeping trees. The subtle curves and bends of the rolling lawns and open terraces, accompanied by unique art pieces dabbled throughout the garden, create an inspiring and enriching experience for anyone who walks through its gates.

The pathways then lead visitors through the Scholar's Walk to the Scholar's Terrace, which overlooks the rich hues of the maple trees in the north garden path. The north garden path moves elegantly through lush planting, and from here, visitors can choose to continue to the Pine Allee, featuring a rest point overlooking Pacific Spirit Park, or pass through the Magnolia Allee. The Magnolia Allee opens out to an expanse of soft green lawn. A Display Garden, sitting across the lawn, provides a vivid canvas for annuals, such as tulips during the spring and begonias during the summer. Each year, a different plant palette is explored as a striking visual living painting that can be enjoyed from the overlooking entertainment terrace. Seven large classical planters, contrasting the broken flagstone pathway set in the lawn, frame this dramatic walkway.

The owners requested a garden that allowed for creative expression through planting designs and their extensive, and growing, art collection. This garden continues to evolve today with the addition of sculptural pieces, custom furniture, and subtle plant refinements.

Moroccan Touch

Designer: Frederic Francis
Design Company: Francis Landscapes
Location: Doha, Qatar
Site Area: 8,000 m^2

Tropical Paradise

Terrain: Waterfront

A series of theme gardens and natural stone courtyards create a choice of living spaces on the grounds of this Doha villa. Despite difference in character and mood, these areas come together as one. They achieve a fluid continuity which, amplified by the presence of water throughout the landscape, culminates in the glistening waters of the adjoining lagoon.

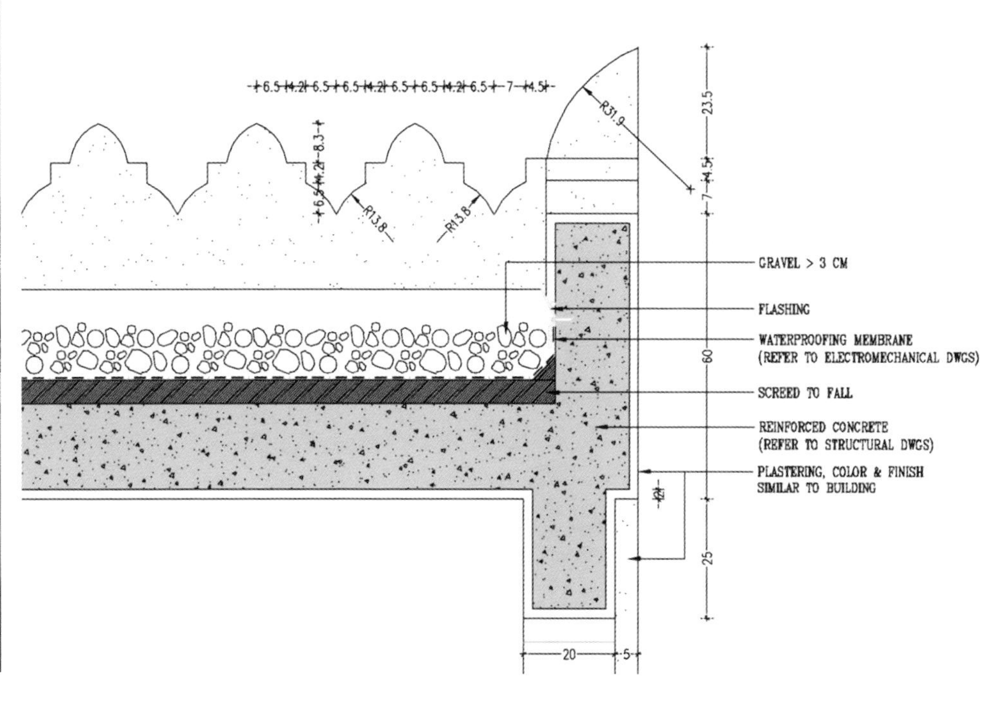

6.5 4.2 6.5 6.5 4.2 6.5 6.5 4.2 6.5 7 4.5
R31.9
23.5
8.3
6.5 4.2
R13.8
R13.8
4.5
7
GRAVEL > 3 CM
FLASHING
WATERPROOFING MEMBRANE
(REFER TO ELECTROMECHANICAL DWGS)
60
SCREED TO FALL
REINFORCED CONCRETE
(REFER TO STRUCTURAL DWGS)
PLASTERING, COLOR & FINISH
SIMILAR TO BUILDING
25
20
5

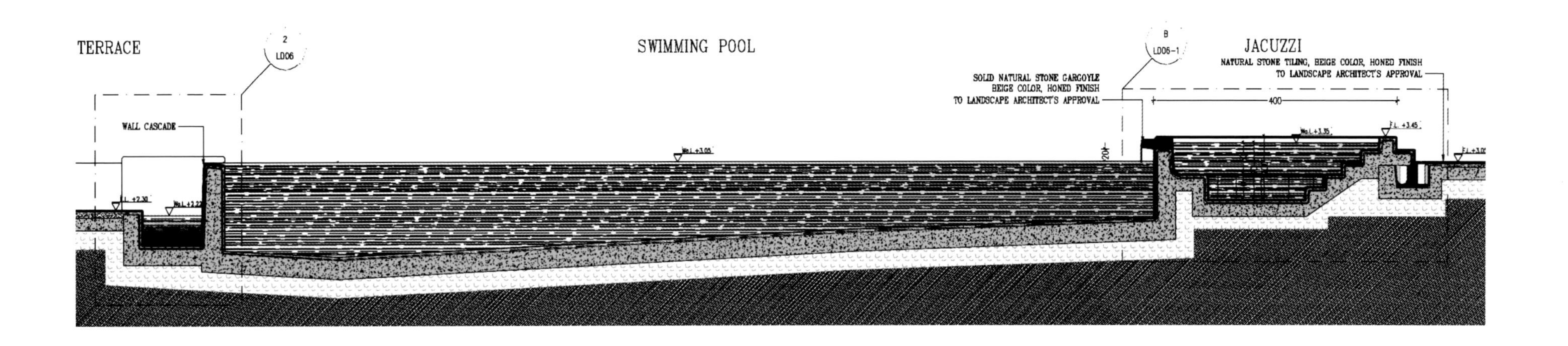

TERRACE
SWIMMING POOL
JACUZZI
2
LD06
WALL CASCADE
B
LD06-1
SOLID NATURAL STONE GARGOYLE
BEIGE COLOR, HONED FINISH
TO LANDSCAPE ARCHITECT'S APPROVAL
NATURAL STONE TILING, BEIGE COLOR, HONED FINISH
TO LANDSCAPE ARCHITECT'S APPROVAL
400

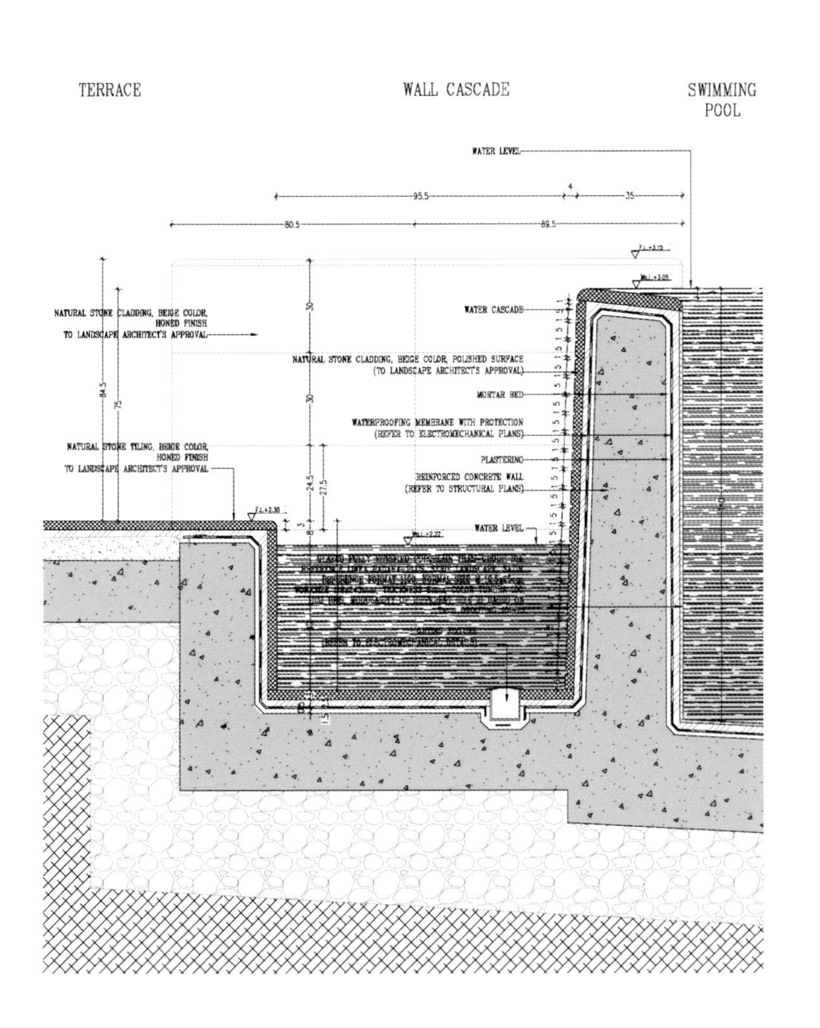

TERRACE
WALL CASCADE
SWIMMING
POOL
WATER LEVEL
95.5
4
35
80.5
89.5
NATURAL STONE CLADDING, BEIGE COLOR,
HONED FINISH
TO LANDSCAPE ARCHITECT'S APPROVAL
WATER CASCADE
NATURAL STONE CLADDING, BEIGE COLOR, POLISHED SURFACE
(TO LANDSCAPE ARCHITECT'S APPROVAL)
MORTAR BED
WATERPROOFING MEMEBRANE WITH PROTECTION
(REFER TO ELECTROMECHANICAL PLANS)
PLASTERING
REINFORCED CONCRETE WALL
(REFER TO STRUCTURAL PLANS)
NATURAL STONE TILING, BEIGE COLOR,
HONED FINISH
TO LANDSCAPE ARCHITECT'S APPROVAL
WATER LEVEL
84.5
75
30
30
24.5
27.5

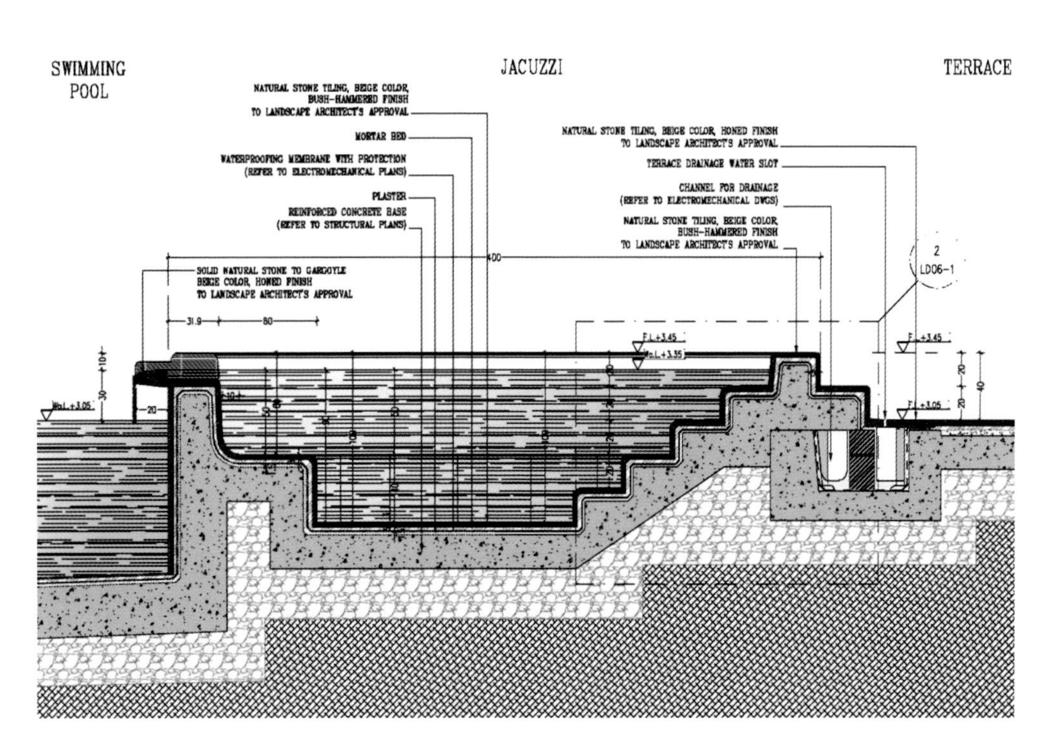

SWIMMING
POOL
JACUZZI
TERRACE
NATURAL STONE TILING, BEIGE COLOR,
BUSH-HAMMERED FINISH
TO LANDSCAPE ARCHITECT'S APPROVAL
MORTAR BED
WATERPROOFING MEMBRANE WITH PROTECTION
(REFER TO ELECTROMECHANICAL PLANS)
PLASTER
REINFORCED CONCRETE BASE
(REFER TO STRUCTURAL PLANS)
NATURAL STONE TILING, BEIGE COLOR, HONED FINISH
TO LANDSCAPE ARCHITECT'S APPROVAL
TERRACE DRAINAGE WATER SLOT
CHANNEL FOR DRAINAGE
(REFER TO ELECTROMECHANICAL DWGS)
NATURAL STONE TILING, BEIGE COLOR,
BUSH-HAMMERED FINISH
TO LANDSCAPE ARCHITECT'S APPROVAL
400
2
LD06-1
SOLID NATURAL STONE TO GARGOYLE
BEIGE COLOR, HONED FINISH
TO LANDSCAPE ARCHITECT'S APPROVAL
31.9
80

Through their pronounced linearity and symmetry, the gardens also complement the Moroccan style of the villa. As a result, a rare balance is achieved between the interior and exterior architectures. This orderly, geometric composition is mirrored by the shapes given to various elements throughout the outdoors: a planter in the form of a square here, a star-shaped fountain there.

At the centre of it all, facing the lagoon, the main water feature unfolds over three levels. Sitting at the top, a large, circular jacuzzi overflows into the swimming pool. From the pool, one can enjoy a unique vantage point of the lagoon, through the framed view offered by a wooden pergola. The impressive rectangle of blue that forms the no-edge pool ends in a cascade collected in a long, narrow channel beneath it. On the same level is a charming, partially shaded Andalusian water channel with an octagonal fountain at the end. Rows of palms stand majestic on either side, framing the scenery. Ahead of the steps which lead to the beach, a reflecting pool mirrors the blue of the swimming pool at one time of day and the clear waters of the lagoon at another.

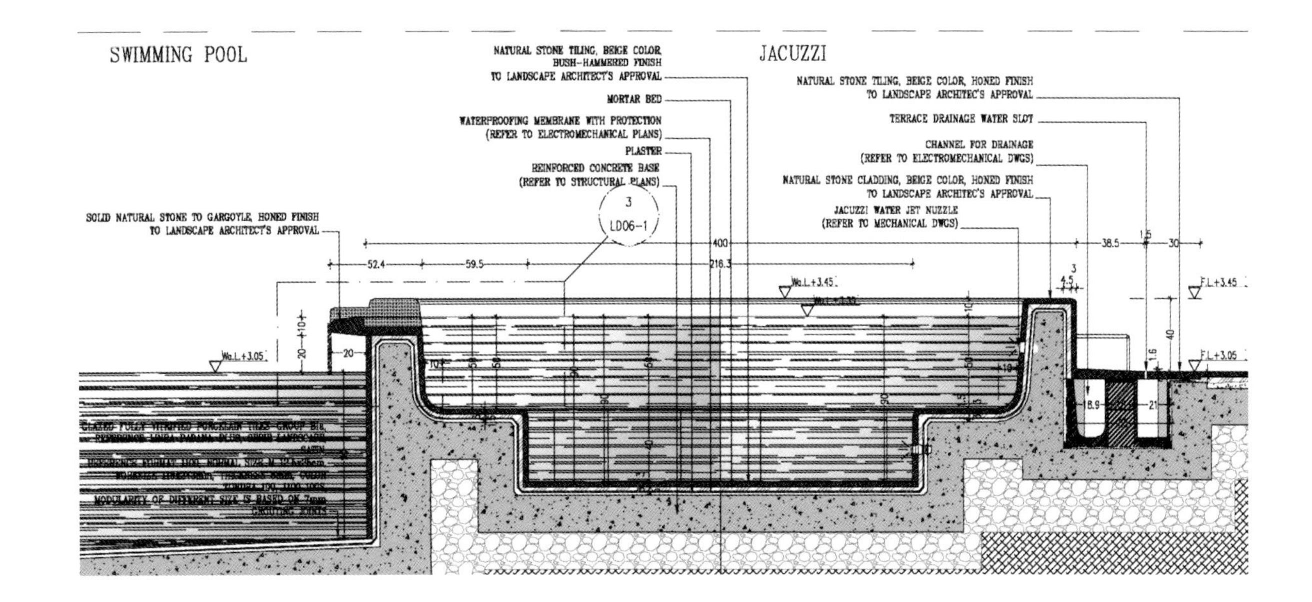

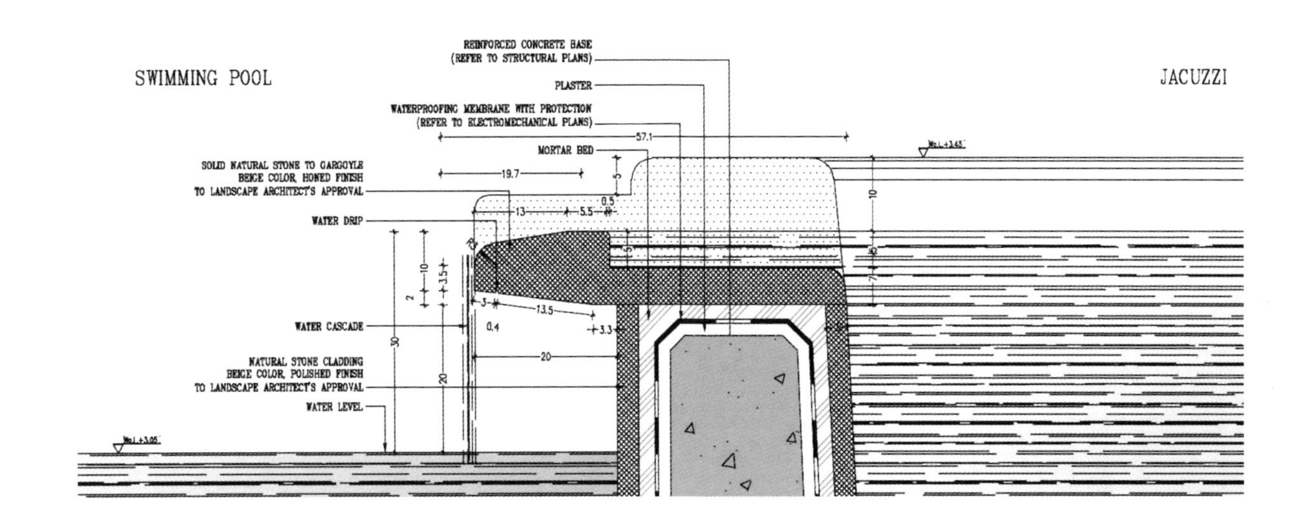

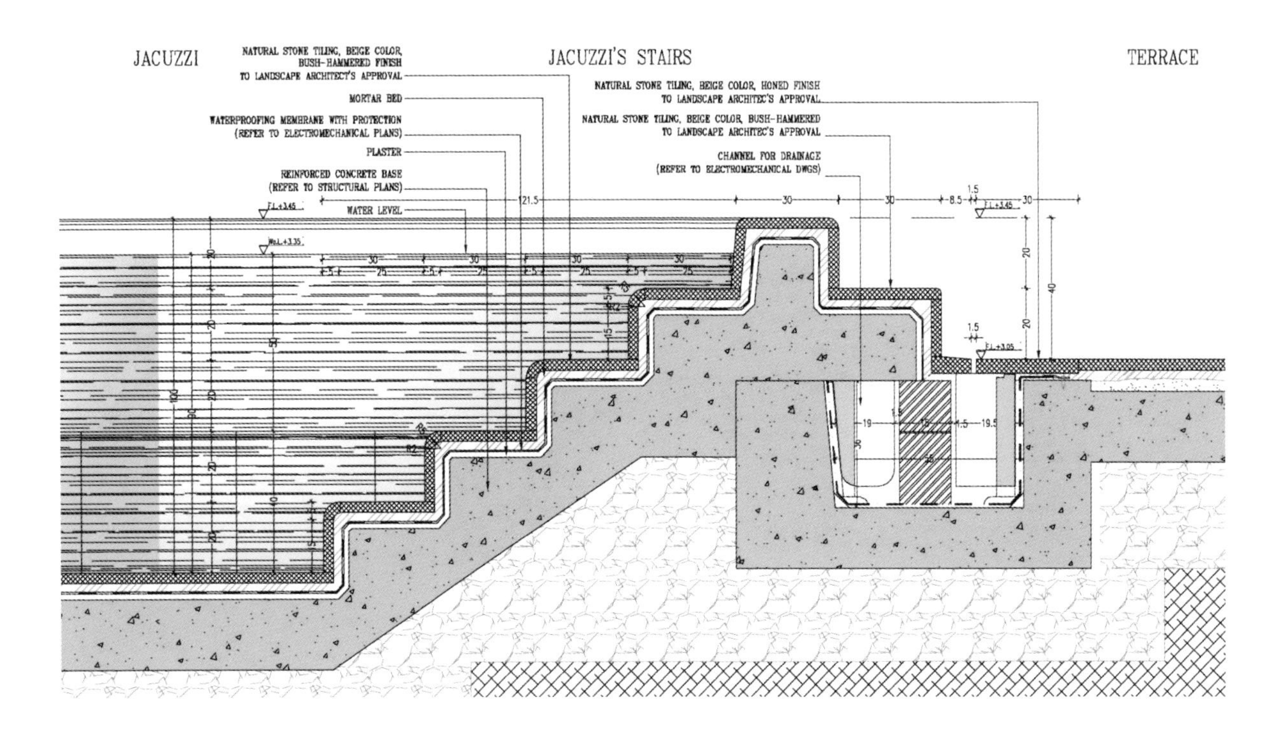

Although the deep blue of the pool and green palette of the trees dominate the landscape, a closer look reveals defiant touches of colour in strategic spots. A delicate wreath of red flowers at the foot of an age-old olive tree, the pink blossoms of a chorisia at the end of a lane of palms, bougainvilleas in full bloom...

The property may owe its richness to the way the big picture fits together, but it also owes it to the smaller details. As you make your way through the gardens, you are sure to notice how impossible it is to take everything in at once. There is always something new to catch the eye: the delicate stone frame of a fountain spout, the mysteries of the secret garden or some other delight waiting to be discovered.

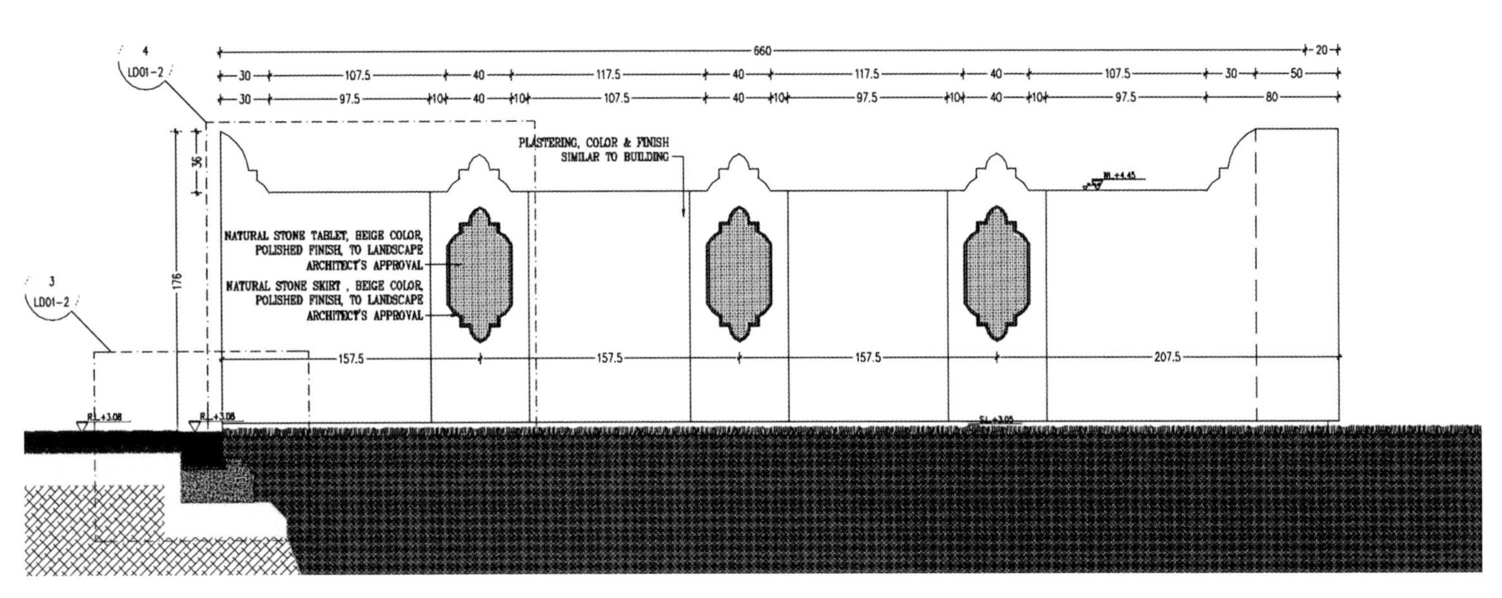

4
LD01-2
3
LD01-2
660
20
30
107.5
40
117.5
40
117.5
40
107.5
30
50
30
97.5
10
40
10
107.5
40
10
97.5
10
40
10
97.5
80
36
176
PLASTERING, COLOR & FINISH
SIMILAR TO BUILDING
NATURAL STONE TABLET, BEIGE COLOR,
POLISHED FINISH, TO LANDSCAPE
ARCHITECT'S APPROVAL
NATURAL STONE SKIRT , BEIGE COLOR,
POLISHED FINISH, TO LANDSCAPE
ARCHITECT'S APPROVAL
157.5
157.5
157.5
207.5
RL +3.08
RL +3.08

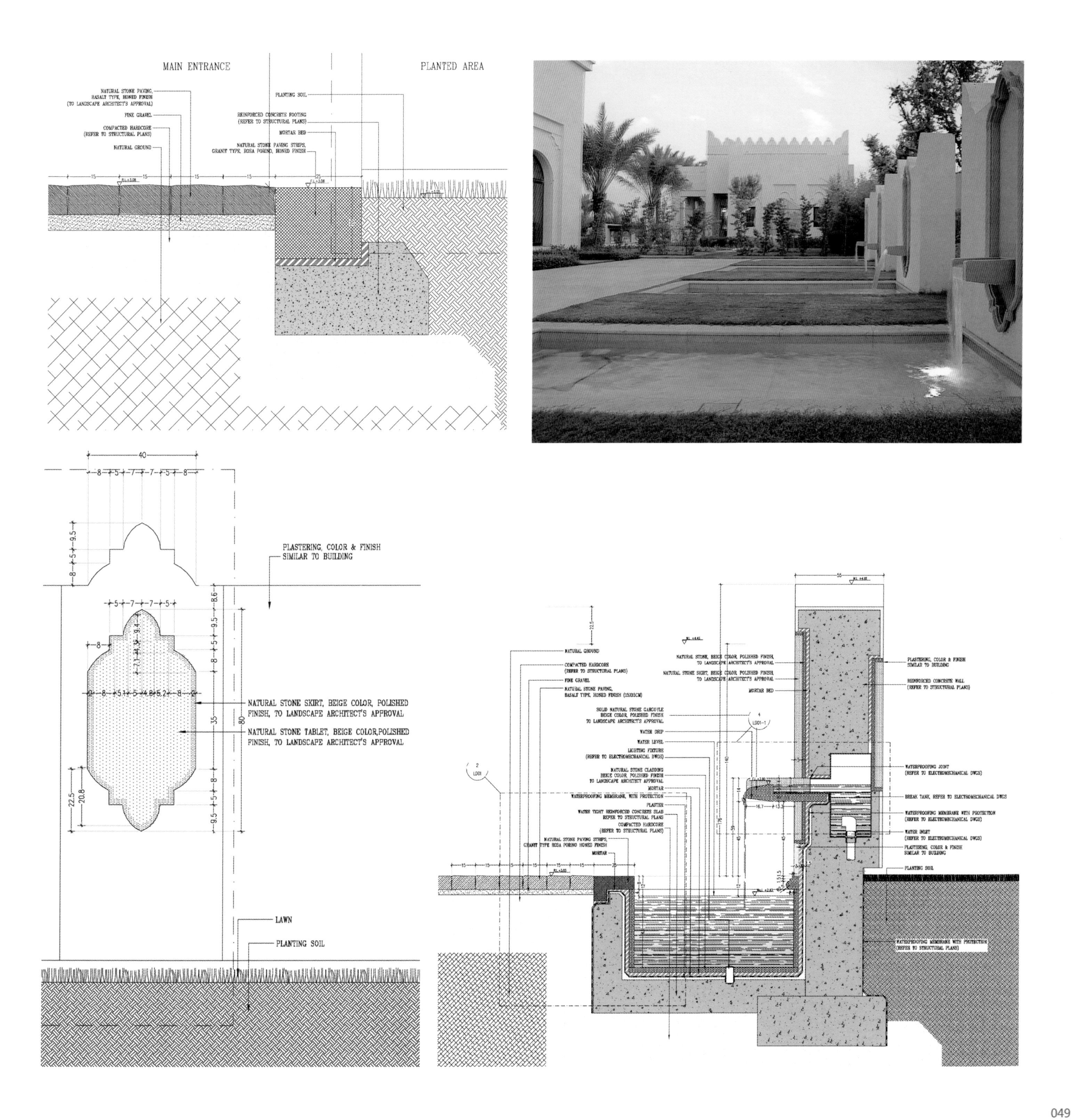

MAIN ENTRANCE
PLANTED AREA
NATURAL STONE PAVING, HASALT TYPE, HONED FINISH (TO LANDSCAPE ARCHITECT'S APPROVAL)
FINE GRAVEL
COMPACTED HARDCORE (REFER TO STRUCTURAL PLANS)
NATURAL GROUND
PLANTING SOIL
REINFORCED CONCRETE FOOTING (REFER TO STRUCTURAL PLANS)
MORTAR BED
NATURAL STONE PAVING STRIPS, GRANIT TYPE, ROSA PORINO, HONED FINISH
PLASTERING, COLOR & FINISH SIMILAR TO BUILDING
NATURAL STONE SKIRT, BEIGE COLOR, POLISHED FINISH, TO LANDSCAPE ARCHITECT'S APPROVAL
NATURAL STONE TABLET, BEIGE COLOR, POLISHED FINISH, TO LANDSCAPE ARCHITECT'S APPROVAL
LAWN
PLANTING SOIL
NATURAL GROUND
COMPACTED HARDCORE (REFER TO STRUCTURAL PLANS)
FINE GRAVEL
NATURAL STONE PAVING, BASALT TYPE, HONED FINISH (15X15CM)
NATURAL STONE, BEIGE COLOR, POLISHED FINISH, TO LANDSCAPE ARCHITECT'S APPROVAL
NATURAL STONE SKIRT, BEIGE COLOR, POLISHED FINISH, TO LANDSCAPE ARCHITECT'S APPROVAL
MORTAR BED
SOLID NATURAL STONE GARGOYLE BEIGE COLOR, POLISHED FINISH TO LANDSCAPE ARCHITECT'S APPROVAL
WATER DRIP
WATER LEVEL
LIGHTING FIXTURE (REFER TO ELECTROMECHANICAL DWGS)
NATURAL STONE CLADDING BEIGE COLOR, POLISHED FINISH TO LANDSCAPE ARCHITECT APPROVAL
MORTAR
WATERPROOFING MEMBRANE, WITH PROTECTION
PLASTER
WATER TIGHT REINFORCED CONCRETE SLAB REFER TO STRUCTURAL PLANS
COMPACTED HARDCORE (REFER TO STRUCTURAL PLANS)
NATURAL STONE PAVING STRIPS, TYPE ROSA PORINO HONED FINISH
MORTAR
PLASTERING, COLOR & FINISH SIMILAR TO BUILDING
REINFORCED CONCRETE WALL (REFER TO STRUCTURAL PLANS)
WATERPROOFING JOINT (REFER TO ELECTROMECHANICAL DWGS)
BREAK TANK, REFER TO ELECTROMECHANICAL DWGS
WATERPROOFING MEMBRANE WITH PROTECTION (REFER TO ELECTROMECHANICAL DWGS)
WATER INLET (REFER TO ELECTROMECHANICAL DWGS)
PLASTERING, COLOR & FINISH SIMILAR TO BUILDING
PLANTING SOIL
WATERPROOFING MEMBRANE WITH PROTECTION (REFER TO STRUCTURAL PLANS)

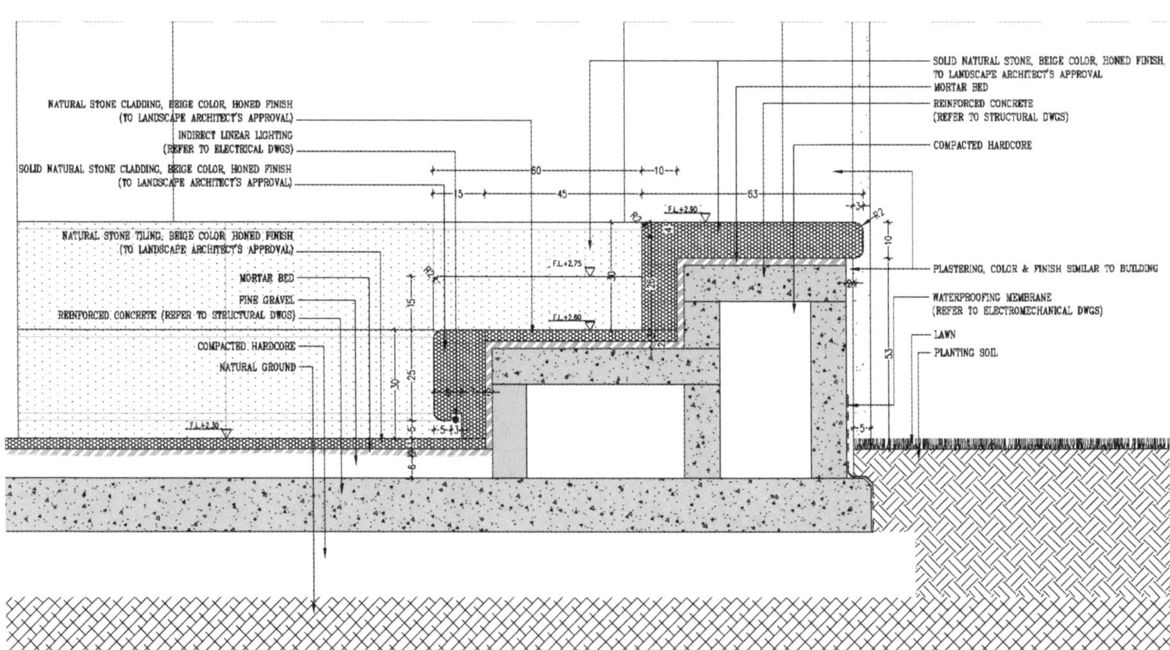

NATURAL STONE CLADDING, BEIGE COLOR, HONED FINISH
(TO LANDSCAPE ARCHITECT'S APPROVAL)
INDIRECT LINEAR LIGHTING
(REFER TO ELECTRICAL DWGS)
SOLID NATURAL STONE CLADDING, BEIGE COLOR, HONED FINISH
(TO LANDSCAPE ARCHITECT'S APPROVAL)
NATURAL STONE TILING, BEIGE COLOR, HONED FINISH
(TO LANDSCAPE ARCHITECT'S APPROVAL)
MORTAR BED
FINE GRAVEL
REINFORCED CONCRETE (REFER TO STRUCTURAL DWGS)
COMPACTED HARDCORE
NATURAL GROUND
SOLID NATURAL STONE, BEIGE COLOR, HONED FINISH,
TO LANDSCAPE ARCHITECT'S APPROVAL
MORTAR BED
REINFORCED CONCRETE
(REFER TO STRUCTURAL DWGS)
COMPACTED HARDCORE
PLASTERING, COLOR & FINISH SIMILAR TO BUILDING
WATERPROOFING MEMBRANE
(REFER TO ELECTROMECHANICAL DWGS)
LAWN
PLANTING SOIL

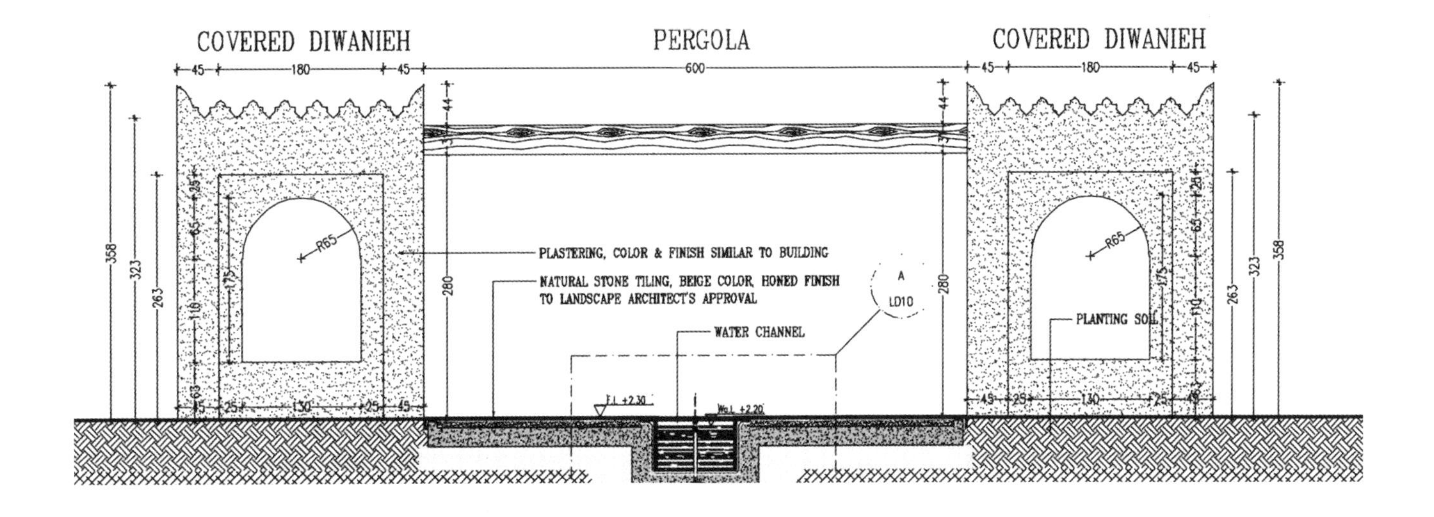
COVERED DIWANIEH
PERGOLA
COVERED DIWANIEH
PLASTERING, COLOR & FINISH SIMILAR TO BUILDING
NATURAL STONE TILING, BEIGE COLOR, HONED FINISH
TO LANDSCAPE ARCHITECT'S APPROVAL
WATER CHANNEL
PLANTING SOIL
A
LD10

Riseley Garden

Design Company: Scott Brown Landscape Design
Location: Melbourne, Victoria, Australia
Site Area: 1,335.5 m^2
Photographer: Patrick Redmond

Formal Elegance

Terrain: Sloping Fields

This water-wise formal garden is a timeless testament to good planning and good design.

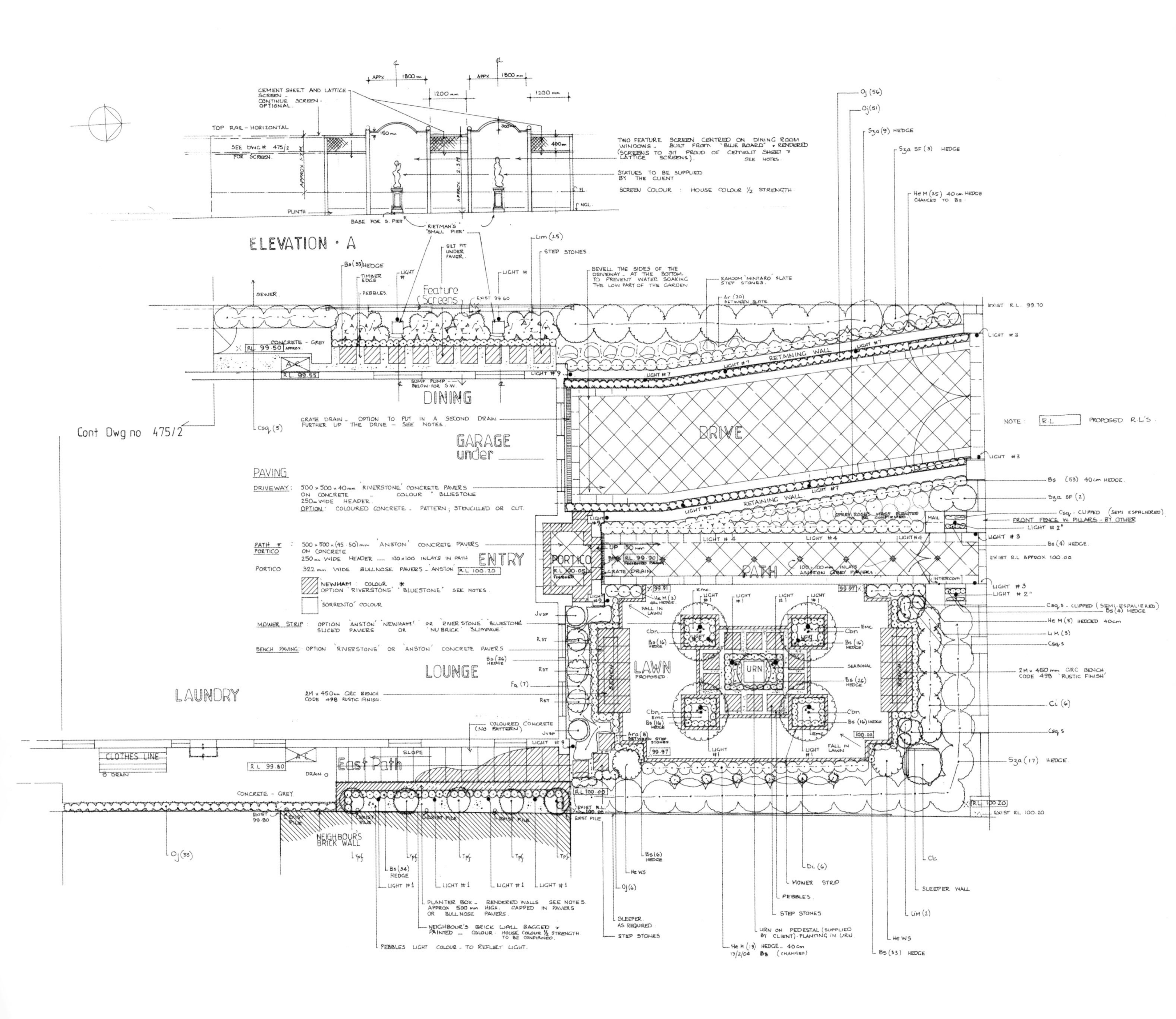

ELEVATION · A
CEMENT SHEET AND LATTICE SCREEN. CONTINUE SCREEN - OPTIONAL.
TOP RAIL - HORIZONTAL
SEE DWG# 475/2 FOR SCREEN.
PLINTH
BASE FOR S. PIER
RIETMAN'S 'SMALL PIER'
TWO FEATURE SCREEN CENTRED ON DINING ROOM WINDOWS - BUILT FROM 'BLUE BOARD' + RENDERED (SCREENS TO SIT PROUD OF CEMENT SHEET + LATTICE SCREENS). SEE NOTES.
STATUES TO BE SUPPLIED BY THE CLIENT
SCREEN COLOUR : HOUSE COLOUR ½ STRENGTH.
Feature Screens
DINING
GARAGE under
DRIVE
ENTRY
PORTICO
PATH
LAWN
URN
LOUNGE
LAUNDRY
East Path
CLOTHES LINE
Cont Dwg no 475/2
PAVING
DRIVEWAY : 500 × 500 × 40mm 'RIVERSTONE' CONCRETE PAVERS ON CONCRETE - COLOUR 'BLUESTONE' 250mm WIDE HEADER.
OPTION : COLOURED CONCRETE - PATTERN, STENCILLED OR CUT.
PATH + PORTICO : 500 × 500 × (45-50)mm 'ANSTON' CONCRETE PAVERS ON CONCRETE 250mm WIDE HEADER - 100×100 INLAYS IN PATH
PORTICO 322 mm WIDE BULLNOSE PAVERS 'ANSTON'
NEWHAM COLOUR OPTION 'RIVERSTONE' 'BLUESTONE' SEE NOTES
'SORRENTO' COLOUR
MOWER STRIP : OPTION 'ANSTON' 'NEWHAM' OR 'RIVERSTONE' 'BLUESTONE' SLICED PAVERS OR 'NUBRICK' 'SLIMPAVE'
BENCH PAVING : OPTION 'RIVERSTONE' OR 'ANSTON' CONCRETE PAVERS
GRATE DRAIN - OPTION TO PUT IN A SECOND DRAIN FURTHER UP THE DRIVE - SEE NOTES.
BEVELL THE SIDES OF THE DRIVEWAY AT THE BOTTOM, TO PREVENT WATER SOAKING THE LOW PART OF THE GARDEN
RANDOM 'MINTARO' SLATE STEP STONES.
RETAINING WALL
NOTE : R.L. PROPOSED R.L'S
FRONT FENCE W. PILLARS - BY OTHER
2M × 450mm GRC BENCH CODE 498 'RUSTIC FINISH'
PLANTER BOX - RENDERED WALLS SEE NOTES. APPROX 500mm HIGH. CAPPED IN PAVERS OR BULLNOSE PAVERS.
NEIGHBOUR'S BRICK WALL BAGGED + PAINTED - COLOUR : HOUSE COLOUR ½ STRENGTH TO BE CONFIRMED.
PEBBLES LIGHT COLOUR - TO REFLECT LIGHT.
NEIGHBOUR'S BRICK WALL
SLEEPER AS REQUIRED
STEP STONES
MOWER STRIP
PEBBLES
URN ON PEDESTAL (SUPPLIED BY CLIENT) - PLANTING IN URN
SLEEPER WALL
COLOURED CONCRETE (NO PATTERN)
CONCRETE - GREY
EXIST R.L. 99.70
EXIST R.L. APPROX 100.00
EXIST RL 100.20

When you enter the front garden you are immediately struck by the elegant formality of the grounds, both of which tie in with the architecture of the home. One of the most critical properties of both the front and rear gardens is they are quite drought-resistant. Considering our current water predicament, this is obviously a much needed prerequisite when designing a new garden, but it's not a feature usually associated with such a formal garden design.

The garage is beneath the house and contains substantial tanks for water run-off and seepage collection. This water storage, combined with the hardy nature of most of the plants, ensure the garden is able to deal with Melbourne's ongoing water restrictions.

The front specimen lawn faces north, which means the entry courtyard, can get quite warm on a sunny day. With this in mind, we used fine leafed Santa Ana couch. This type of turf can develop a high degree of drought resistance and will repair itself should any sections die off during particularly long dry spells.

The other planting consists of repeated hedges, adding to the structural integrity of the garden. These include Dutch box (*Buxus sempervirens* 'Suffruticosa') and rows of *Lirippe* 'Royal Purple', a very drought-resistant plant. *Hebe* 'Wiri Spears' (a New Zealand native), standard roses and four matching specimens of "designer" Indian bean tree (*Catalpa bignonioides*) completes the front courtyard picture. The Indian bean trees surround a large Grecian urn in the middle of the lawn, which gives this small entry courtyard a focal point.

The use of boundary hedges to screen the neighbouring homes and gardens, as well as hide the boundary fences, has given the grounds a lush, timeless appeal. By using native lily pilly hybrids – Syzygium australe in the front and a select form of Acmena smithii in the back – these hedges have contributed to the enduring, classical look of the property, but without the need to introduce into the mix plants with high watering needs.

The back garden contributes a wonderful outlook that can be enjoyed from inside the house all year round, while also catering for the family's every outdoor entertaining requirement. The swim spa, for example, not only provides an intimate experience for one to eight people; it's also a great exercise pool. For maximum convenience, the spa is positioned close to the toilet/bathroom built in the corner of the rear courtyard.

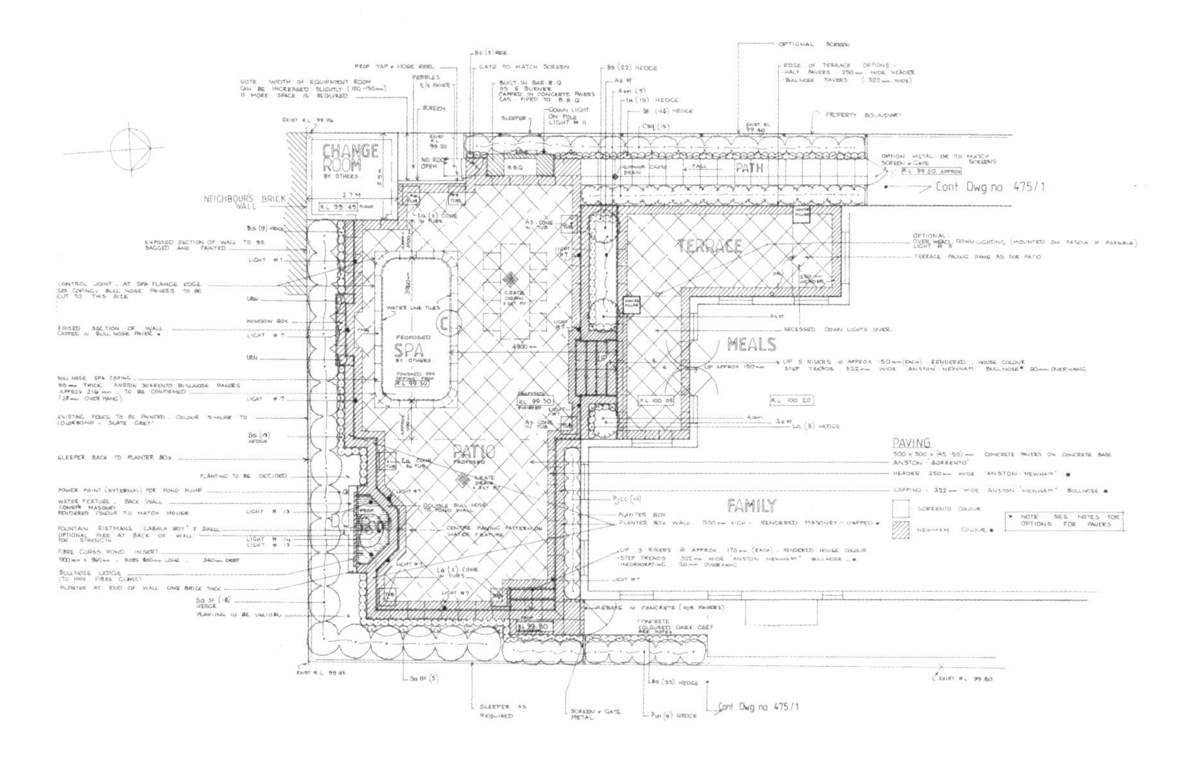

To complete the look in the front garden, we used sand-blasted concrete pavers (from Anston) and appropriate furniture in the form of a garden seat. The dining area is surrounded by the house, a garden area and a water feature-fountain, and is situated to one side of the courtyard. This provides a more intimate dining experience and helps to separate diners from the noise of the spa. At night, lighting ensures this area takes on a whole new dimension.

The patio attached to the house is partially enclosed with a hedge of box leaf privet (*Ligustrum undulatum*), which forms a living balustrade. This continues the theme of partially dividing the rear courtyard into smaller sub-rooms, the result being more interest and more intimacy.

The western side of the home has a hedge of Sasanqua camellias and there are statuettes positioned outside the window of the formal dining and lounge rooms. The lighting and the rendered screen built to form the backdrop provide a stunning outlook both day and night from these rooms of the house.

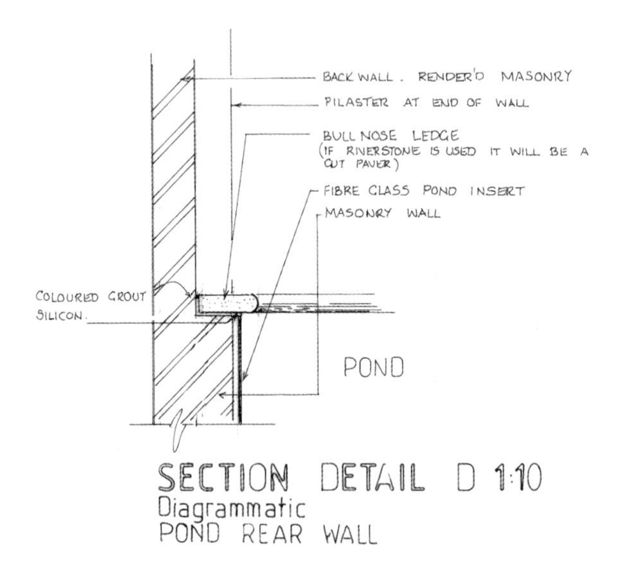

SECTION DETAIL D 1:10
Diagrammatic
POND REAR WALL

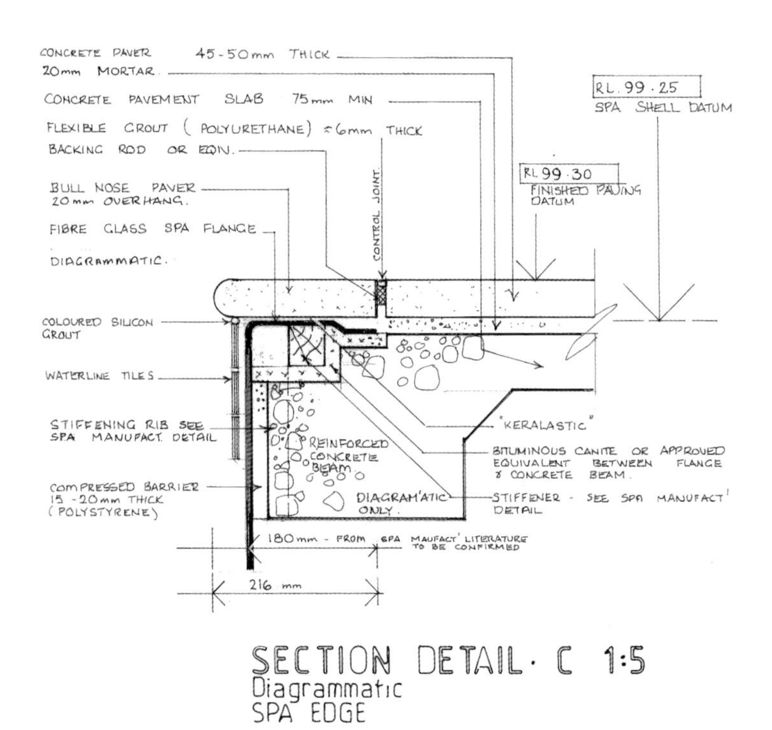

SECTION DETAIL · C 1:5
Diagrammatic
SPA EDGE

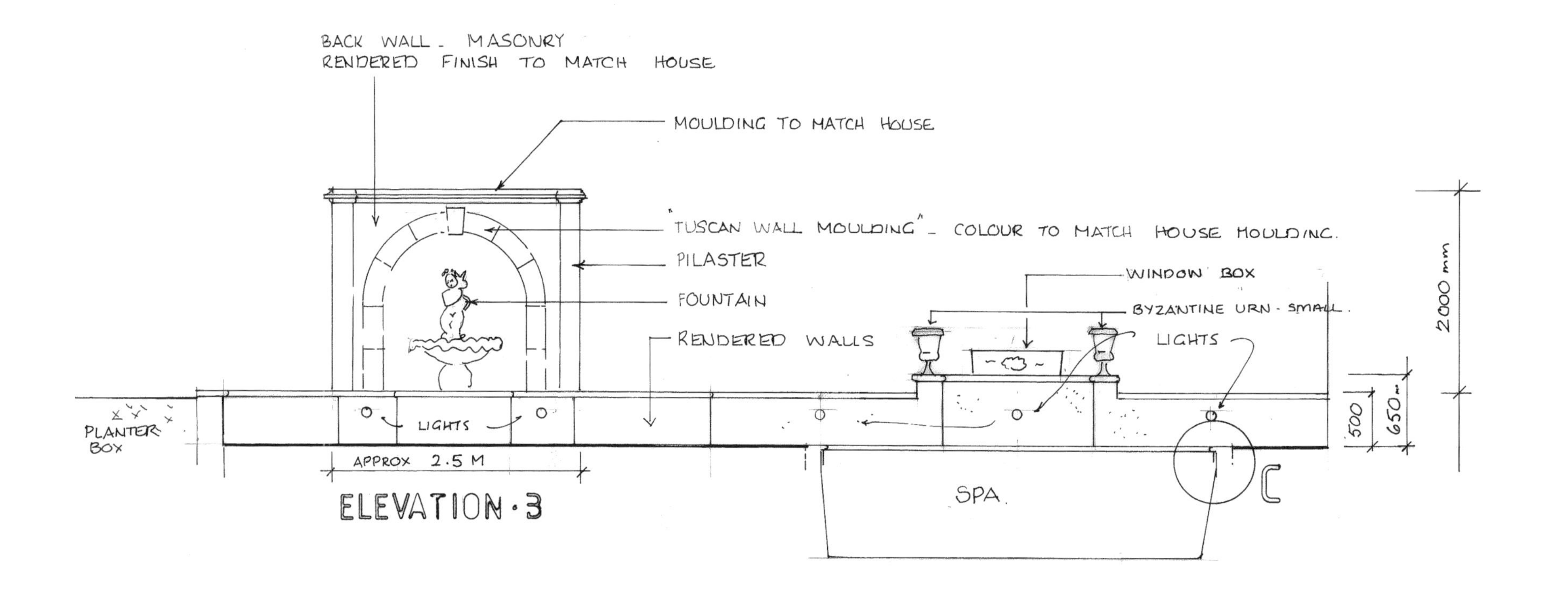

BACK WALL - MASONRY
RENDERED FINISH TO MATCH HOUSE
MOULDING TO MATCH HOUSE
"TUSCAN WALL MOULDING" - COLOUR TO MATCH HOUSE MOULDING.
PILASTER
FOUNTAIN
RENDERED WALLS
WINDOW BOX
BYZANTINE URN - SMALL.
LIGHTS
LIGHTS
PLANTER BOX
APPROX 2.5 M
ELEVATION · 3
SPA
2000 mm
500
650

Donnellan Garden

Design Company: Scott Brown Landscape Design
Location: Melbourne, Victoria, Australia
Site Area: 8,093.7 m^2
Photographer: Patrick Redmond

When European Elegance Meets Natural Wild

Terrain: Mountainous Region

The design brief may have been straightforward – relate the structured elegance of the residence to its wild, natural bushland setting – but reconciling the inherent contradictions this posed required a great deal of thought.

Because the front yard was relatively level, unlike the back garden with its steep natural fall, we were able to implement a rather grand, formal design which imbues the house with a greater "sense of place" when viewed from the street. At the back of the house, however, we needed to resolve the challenges presented by the precipitous fall. This was achieved by creating a "floating" pool and an adjacent entertainment area, both supported by pillars.

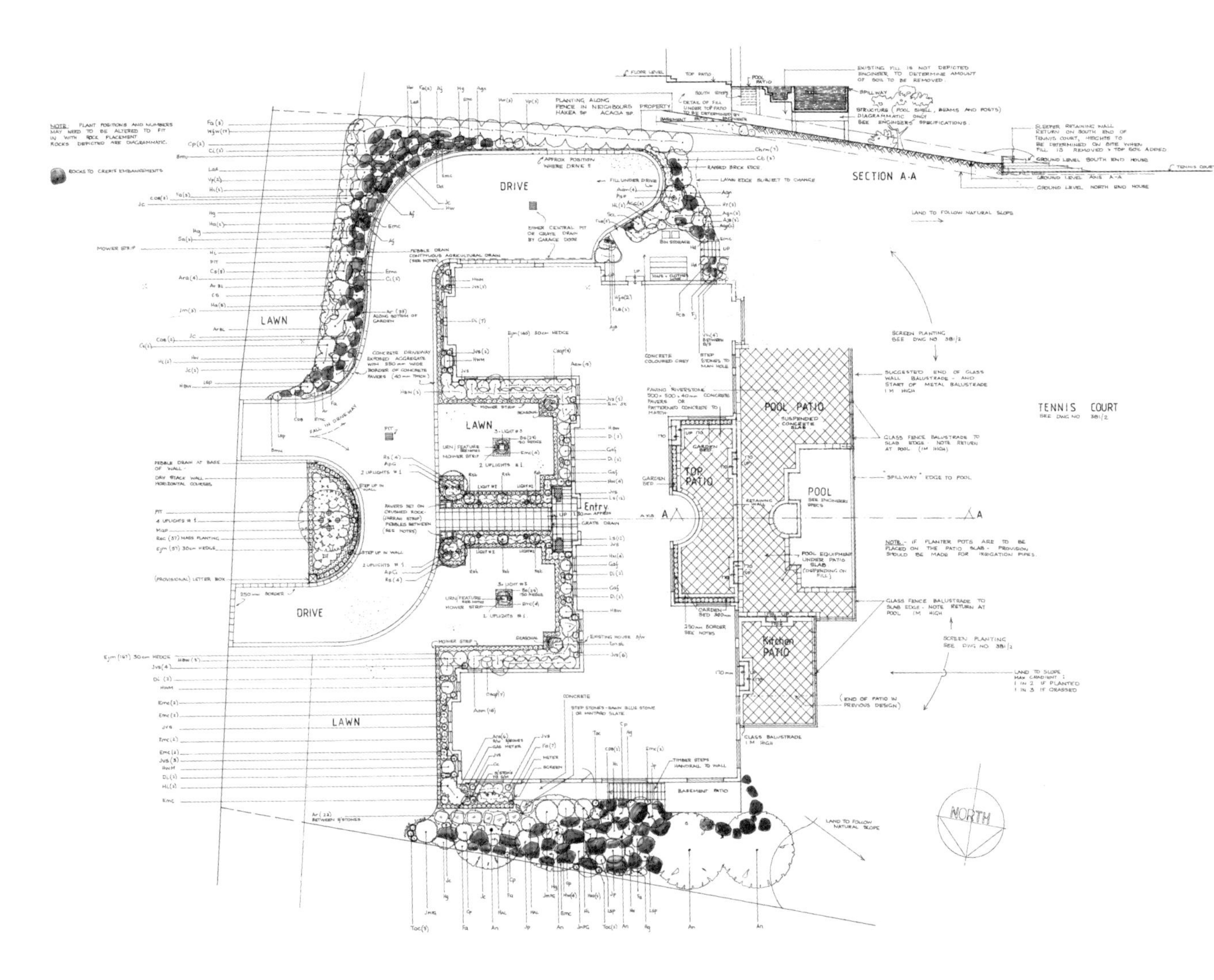

The back garden has been designed to function as a series of structured outdoor "rooms". The sheer, three-four metre drop-off edge of the pool, along with the fenced perimeter of the outdoor dining and lawn/ garden areas, separate the living and entertainment area at the rear of the home from the untamed natural landscape below. This lower part of the block merges with the surrounding countryside, connecting the garden to a larger environment dominated by rolling hills, rocky gullies and clump plantings of manna gums.

To create such an easily accessible and functional living area required significant engineering expertise. To imbue the space with such a sense of luxury required just the right use of materials, including the exposed aggregate pavement, the grey-black pool coping and pool fence posts, the large concrete planters and the wrought iron dining chairs – all of which work together to create a unified look.

In the front yard, access to the five-car garage is via an expansive semi-circular driveway where the customised exposed aggregate surface tones down what might otherwise have been an overly dominant feature. The colours of the stones used in the driveway match the colours expressed in the exterior finishes of the residence – namely charcoal, grey, cream and white. As buff sandstone-coloured concrete pavers were used for the front porch and path, they were also used as a decorative border along the driveway thereby visually linking the key hard landscape elements.

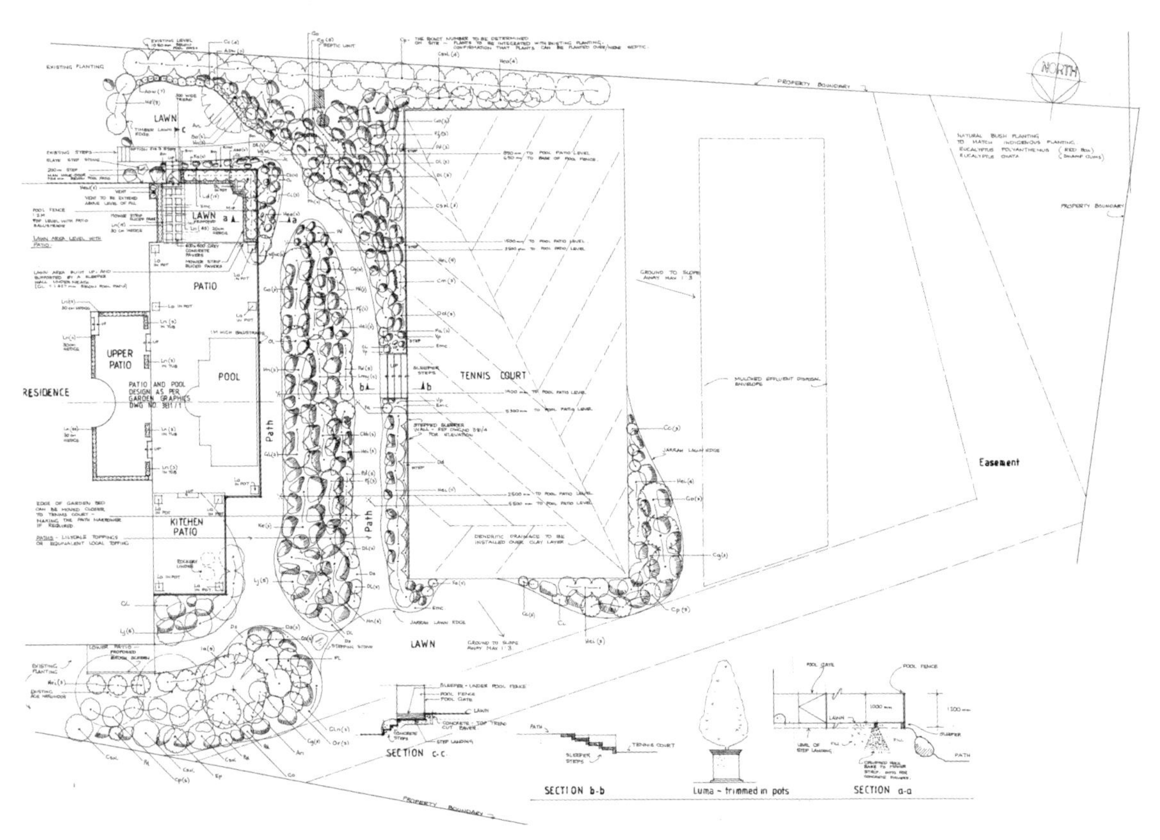

As a further complement to the quiet elegance of the home, the floral colour palette in the front garden is restricted to white, cream, mauve and maroon. Foliage colours vary from dark green to mint green, and from grey-green to blue-green. These tightly foliaged plants – such as the Lonicera hedges are used close to the ground, in keeping with a traditional topiary-style approach. Alternatively, some of the more vertical plantings – like the centrally located weeping mulberry, Indian Bean Tree, Juniper Skyrockets, and dwarf lavender – were selected with the nearby bushland in mind. Their less dense, slightly feathered foliage balances the need for formality with a softer, less manicured ambience. At night, the vertical plantings, along with the two-tiered concrete water features, are uplit, casting a magical glow.

The two fountains play an important role in connecting the garden to the house. Charcoal grey in colour, they complement the tiled roof. This fosters a sense of belonging without running the risk of the fountains being "lost" against the exterior façade of the house, which would have happened if a lighter colour had been used. In addition, the subtle sound of cascading water serves as a link to the pool at the rear, which is visible upon opening the front door.

The outdoor environment has been designed specifically to juggle the contradiction between the classical formal elegance of the home and the "natural", Australian regional setting of the property. This has been successfully achieved by thematically relating the areas adjacent to the home to the house style itself, whilst blending more formal style with the surrounding bushland as one moves further away from the house. The result is that the contradiction has been avoided and the "setting" of the home is "appropriate".

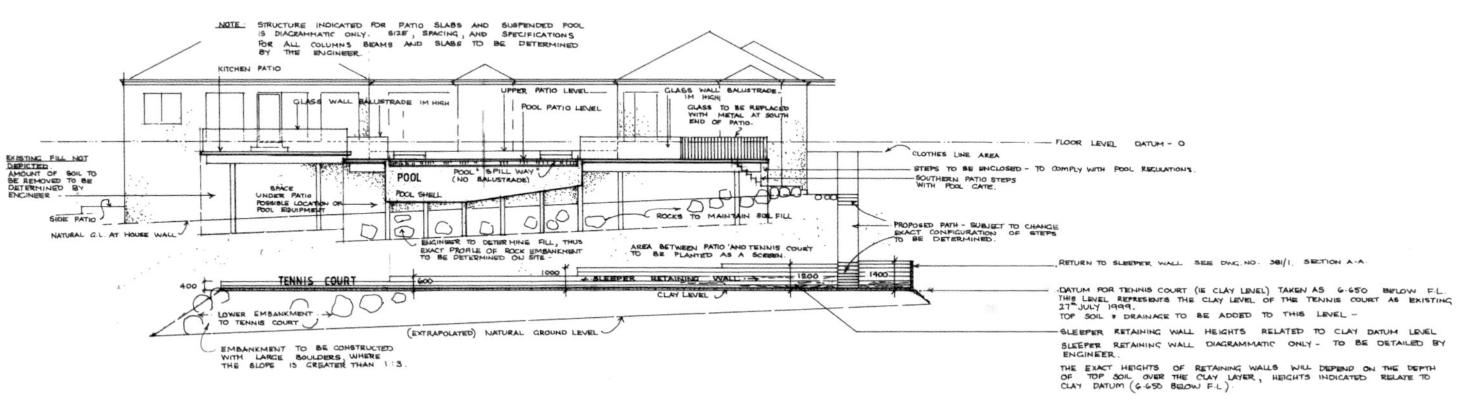

ELEVATION B View westward Tennis court & Pool patio

K House

Design Company: Doyle Herman Design Associates
Location: Greenwich, Connecticut, U.S.A
Site Area: 8,093.7 m^2
Photographer: Esto & Neil Landino

1800 Rustic House

Terrain: Mountainous Region

This traditional rustic house dates back to the early 1800's and is located in Greenwich, Connecticut. After a sympathetic renovation to the home was completed in 2004, a distinctive new landscape was begun in 2005 and continues to this day, which compliments the history and setting of the site and fuses it with the architecture. Set upon eight acres, this property utilizes a series of stone retaining walls to shape the natural sloping land into usable, flat garden spaces. The house sits at the high point of the property with wetlands at the basin. The farm, true to its past, produces organic fruits and vegetables and houses a variety of fowl. The design process has resulted in a dynamic and productive dialogue that intertwines the rustic, natural setting and formalized European landscape with usable exterior spaces and activities on approximately four of the eight acres.

The programmatic intent of the designed landscape was to build upon and clarify the relationships between the bucolic countryside landscape and more formalized gardens as well as to make better use of the sloping site. Careful consideration was given to plant combinations for a structured, yet pared down gesture. Before the installation, care was taken to save and reuse existing plant material, including several mature apple trees. Two types of locally sourced stones, antique granite slabs and reclaimed field stones were used to construct the retaining walls and steps which connect the upper and lower sections of the property.

Descending from the house, the design evolves from formalized spaces, to a rustic landscape that echoes the history of the site. The Taxus capitata allée acts a connecting space both aesthetically and physically.

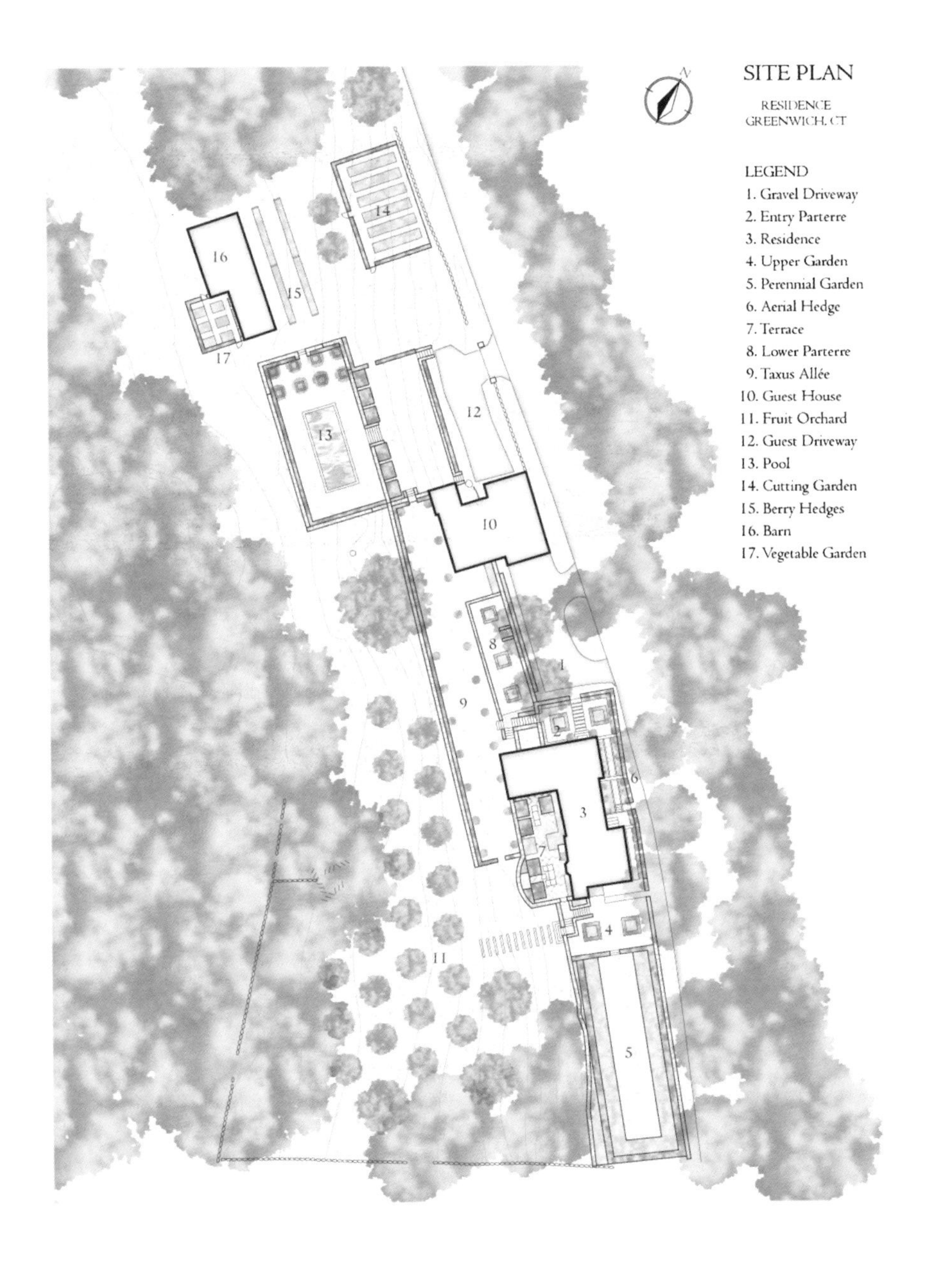

The entry parterre is defined and enclosed by an aerial hedge and a low masonry wall. Sheared square plantings of boxwood and hornbeams relate to the entrance points of the residence and set up clean lines that connect the surrounding spaces. The use of cream colored gravel pathways brighten the entrance parterre, promotes natural drainage and contrast the boldness of the antique granite slab pathway.

On axis with the house, the perennial garden emphasizes understated cool tones of plants. The perennial garden is enclosed with a tall *Carpinus betulus* hedge. Directly off the back of the house, the permeable brick patio is separated into dining and entertaining spaces by low hedges and aerial hedges of *Tilia cordata*. Custom made reclaimed teak sofas, coffee table and side serving tables and a custom stone fireplace complete the space.

A series of two smaller stone retaining walls connect the lower Taxus allée from the entry parterre. The composition of the *Taxus capitata* allée provides both a visual and physical experience as it has direct access from both sides of the house and is also the main passageway between the top and bottom of the property. The repeating form of sheared square plantings of boxwood and hornbeams act as a transitional element between the elevations. Hornbeam and beech hedges are continuously utilized to establish a degree of enclosure in all of the landscape spaces. Drought tolerant fescue grass with the addition of Asclepias verticillata plants act to soften the sheared plantings and showcase the mixture of rustic and traditional landscape.

Another key component of the site is the agrarian landscape elements which add scenery to the property and biodiversity. A fruit orchard laid out in a diagonal grid parallel to the slope consists of old, antique English and French specimens of plum, apple and apricot trees. Additionally, a vegetable and cutting garden and rows of raspberry and blackberry give the clients a link to the past history of the property.

Continuity of style and materials in the landscape can be clearly seen in the landscape elements throughout the levels of the property. The reappearance of materials and patterns throughout the landscape ensures that the mixed outdoor spaces flow from one to another seamlessly and harmoniously.

Old Mill Farm

Design Company: Doyle Herman Design Associates
Location: Greenwich, Connecticut, U.S.A
Site Area: 303,514.2 m^2
Photographer: Marion Brenner & Neli Landino

English Style Garden

Terrain: Flat Ground

Old Mill Farm is one of the last great estates of Greenwich, Connecticut which is comprised of seventy-five +/- acres in the back country of Greenwich. Designed by noted architect Charles Lewis Bowman for prominent financier George Lewis Ohrstrom, the design was awarded a medal by the Greenwich Board of Trade in 1931 and has also been featured in numerous architectural and historical publications. While the home was built between 1926 and 1927, the gardens are a more modern addition to the Elizabethan-inspired Tudor manse.

The current client became the third owner of the property in 1994 and the landscape designer was hired that same year to design and oversee a new era in the estate's history. It was the client's respect for the history of the house and the surrounding grounds that set the tone for the relationship between the client and the landscape designer. None of the original gardens were evident, except for neglected garden patches and poorly maintained masonry walls and walkways. It was agreed to make the best of these; restore original hardscapings, and first and foremost, compliment the architectural style of the home. It was also instructed that the Elizabethan-inspired Tudor manse was to have a traditional landscape, English in style. The last element in the comprehensive site plan was completed in 2006 and over the 12 year course of work on the 15 acres of gardens.

LEGEND

1. Water Fountains
2. Topiary Garden
3. Residence
4. Parterre Garden
5. Orchard
7. Pool
8. Patio
9. Perennial Garden
11. Tennis Court
12. Vegetable Garden
13. Rose Garden
14. Limestone Folly

SITE PLAN

RESIDENCE
GREENWICH, CT

Starting at the front of the house, the central water feature with fountains, constructed in 1999, is flanked on either side by three Fagus sylvatica "Fastigiata". Opposite this is a topiary garden of traditional yew shapes, some of which are as old as the home itself. These were sourced and restored by the designer and newer ones have been created and added. To the east of the front door, one is also given a hint of its private gardens through a wrought iron gate and a wall pierced with a grille of the same style. Through the gate it is clear to see that these gardens maintain a tradition of rich planting within a strong framework of garden "rooms". These include a double herbaceous border, a rose garden, a parterre garden and a chess set garden, with each being private behind stone walls and tall hedges. An enclosed parterre of Buxus sempervirens "Suffruticosa" with brick pathways sits in half shade. Restoration on these pathways began in 1996 and they are now back to their original condition with a moss coating adding period charm. This is annually planted with Coleus "Inky Fingers" and the fringes of this garden are softened with a fine variety of shade tolerant plants.

Strong structured plantings are a continuous theme but flowers are not neglected on the property. Passing an intimate seating area near the Red Brick Garden, the double herbaceous border "room" is entered. These borders are on either side of a grass walk backed by green and purple hedges of Fagus sylvatica "Dawyck". Above this garden, a high south-facing wall is a backdrop for a rose collection flowering mainly in colors from pink to crimson.

Continuing through the gardens, an atmosphere of quiet but beautiful seclusion is retained. The whole only gradually unfolds as the garden is explored. One walks through another narrow, intimate, double border beside the house, whose bluestone pathway is lined with a mass of lime-green flowers of Alchemilla mollis. This acts as the edging of the borders which contain a mixture of herbaceous plants, and is a lovely foil to the more architectural elements such as small box topiaries and imported staddlestones. This pathway also overlooks the chess set which was commissioned by the designer in 1998, made from fiberglass but finished to mimic stone. This garden is enclosed by a castellated Taxus hedge and a sheared aerial hedge of Carpinus betulus. Openings in these hornbeams reveal a view of an orchard and a distant pond.

The orchard, installed in 2002, consists of thirty-two mature trees in a grid pattern, yielding apples, pears, peaches, plums, cherries and apricots. These were hand picked from a New York state fruit grower and transplanted one at a time to a sloped hill near the house. On a cross axis with this installation, a new garden folly, built in 2004, of limestone with a lead coated copper dome, terminates the view. This imposing feature now sits amongst a new collection of antique apples tree varieties installed in 2008. One can also enjoy a glimpse of this temple rotunda as you meander through the gardens.

The Tudor architectural details are nowhere more evident than at the rear of the house. The rear terrace plantings provide strong structure throughout the year but allow for spectacular seasonal plantings that compliment this great façade. The views here are majestically picturesque as there are areas of open sunlit lawn, in contrast to more formal gardens below the terrace. A set of steps lead you through rose clad arches to a swimming pool and more planned borders with a charming birdhouse as a focal point. The pool area and associated gardens were constructed in 1997. The swimming pool, originally built in 1963, was renovated and enclosed with a wall and wrought iron fencing. This area is simply planted with two sets of four cubed Platanus x acerifolia "Bloodgood". Each underplanted with sheared Buxus sempervirens "boots"; they form a natural shelter from the summer sun. The gates and iron railings were designed to be very much in keeping with the style of the home and are an important element in the overall design.

And finally, another set of steps directs you through the Hydrangea Walk towards the kitchen garden (1998), the tennis court and the Taxus Maze. The formal kitchen and cut flower garden is carefully laid out with narrow beds in a pattern that allows for easy maintenance and harvesting. The orderliness of this layout is enclosed by a tall and dense

Privet hedge. This garden is backed by a pavilion and pergola which services the tennis court. The designer was inspired to add the pavilion to the site design in 2000 when the client asked him to find a use for old limestone window surrounds that they had purchased. Shade is provided with climbing roses, Concord grapes and honeysuckles on the stone and rough sawn cedar pergola. The maze, with its Cordyline potted center, is made of six foot tall Taxus, and is bordered by another purple Fagus sylvatica hedge. The maze garden, installed in 1999, has reached full maturity and provides the client with yet another whimsical pleasure.

Green Leisure Estate

Terrain: Mountainous Region

The customers, an elder couple, together with their family, have resided in a farmhouse for many years. A part of this farmhouse (the former stable) was recently converted for the couple into a modern, ground level, living area. The adjacent garden (ca.270 m²) was to be completely redesigned. A terrace for about 10 people, a pergola for shade, a not too big lawn and attractive plants were all requested. The customer also wished for some protection against the sometimes fierce Western Winds. A second seating area for only two or three people was conceivable. It was their special wish to plant a big larch (*Larix decidua*) in the southern part of the garden. The cubic shaped, glassed-in entrance area was also to be designed accordingly. It was to be planted attractively as the adjacent yard area had no green areas.

Garden Floegel/Kuhr

Designer: Alexander Oesterheld
Location: Dinklar, Germany
Site Area: 270 m²
Photographer: Ferdinand Graf von Luckner

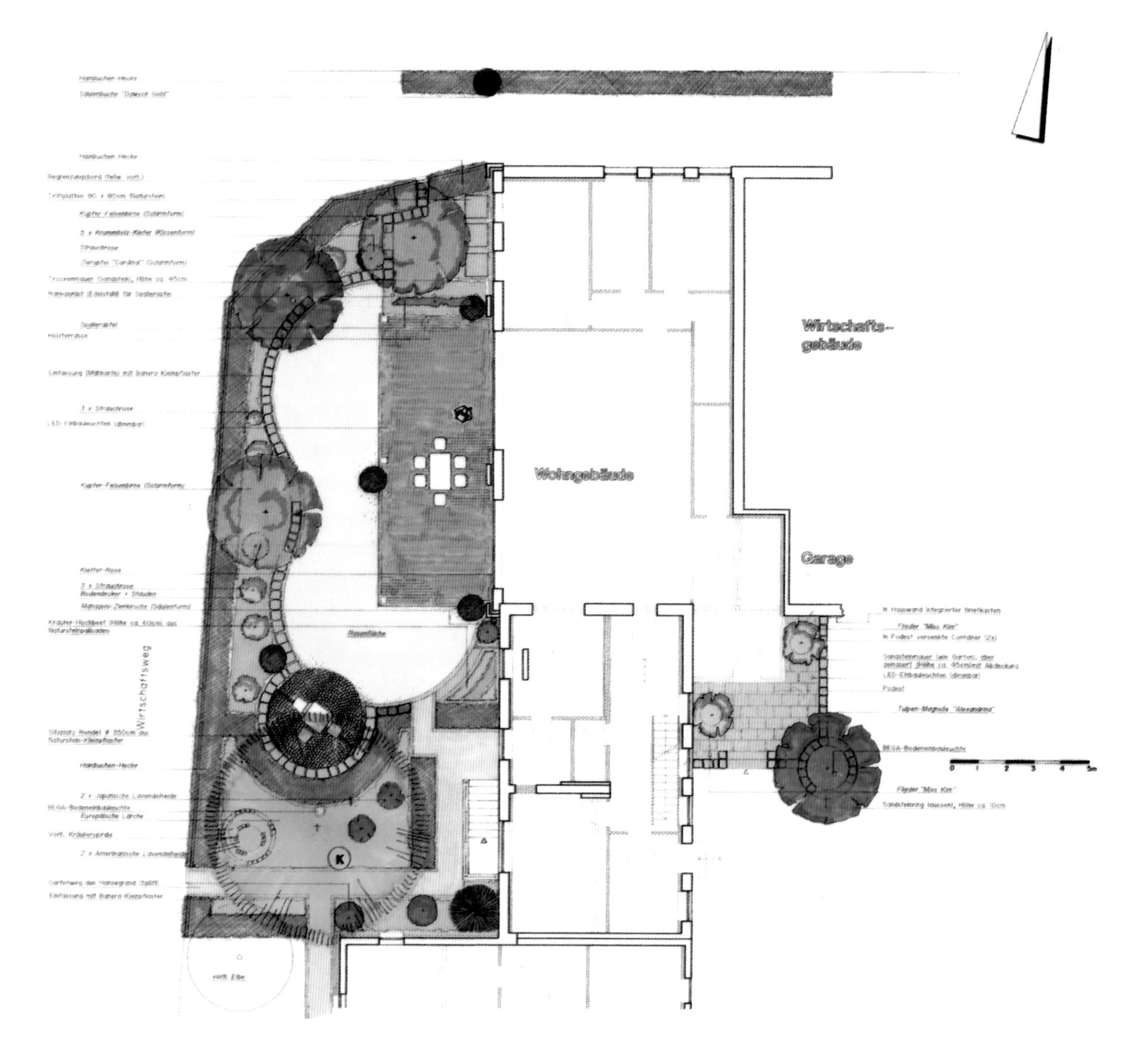

A large wooden terrace (5 m x 10 m) was constructed adjacent to the converted living area. A pergola was put up to provide shade to the seating area. To compensate for the property sloping to the west, a natural stone-wall made of sandstone was erected next to the lawn to terrace off the property. A second seating area, with a round seating platform built on a bed of small granite cobblestones, was constructed in the southern part of the garden. A large parasol was placed there as well to provide the necessary shade. The paths in the southern part of the garden were covered with gravel stones. A raised herb garden (near to the kitchen) was laid out directly to the south of the terrace. A stainless steel plant frame was put right to its north to offer a place for an espalier apple tree. On the raised plant beds to the west, some ornamental trees as well as roses, shrubs and perennials were planted. By special request of the customer, a large Larch (*Larix decidua*) was planted in the southern part of the garden. A hornbeam hedge was planted on the western border of the garden to provide the garden with visual and weather protection. For the purpose of simplifying the maintenance of the garden, an automatic watering system was installed throughout.

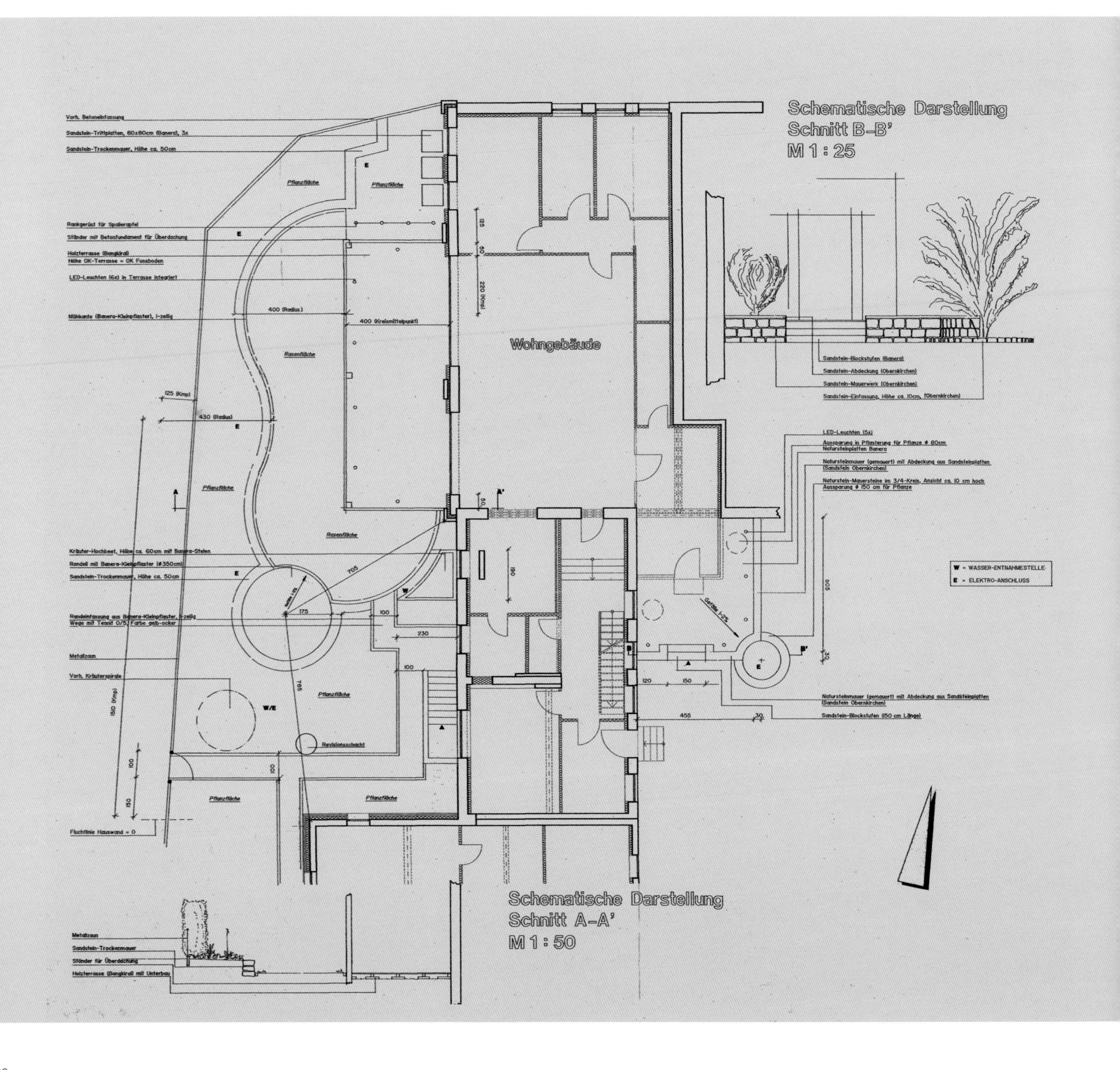
Schematische Darstellung
Schnitt B-B'
M 1 : 25
Wohngebäude
Schematische Darstellung
Schnitt A-A'
M 1 : 50
W = WASSER-ENTNAHMESTELLE
E = ELEKTRO-ANSCHLUSS

City Oasis

Terrain: Flat Ground

Dr. and Mrs. Fuleihan Residence

Designer: Mariane Wheatley-Miller
Design Company: A J Miller Landscape Architecture PLLC
Location: New York, U.S.A.
Site Area: 262 m^2
Photographer: Charles Wright

The concept of this project was to create a patio garden that had a flavour of the Lebanon and Greece. The area is used frequently for"alfresco dining", swimming and entertaining with family and friends.

As the garden area is only a small space and overlooked by adjoining properties we created some privacy with fencing yet created an open and bright feeling. The existing brick paving were removed and replaced with a light beige Travertine. The lighter color reflects the clients eclectic collects and colorful blue ceramic planters, the pool and many colorful plants.

As designers we were also instrumental in advising our client to repaint the existing home from white and grey to a warm beige with green trim highlights, a new brown roof and copper gutters. This "anchored" the house to the site and complemented the hard landscape materials, reducing glare, helping with the scale of the project and creating a warmer more inviting atmosphere. We also provided the six large fibre glass plant containers for the project and undertook to regularly create dramatic seasonal plant displays.

The vinel construction swimming pool existed with the steel edging exposed. We can tervlierev the new paver over the edge to hide the trim and provide a shadow line just over the water. On the axis of the pool and sunroom we positioned a pergola sitting area with gas fireplace and lighting for entertaining. A new perimeter fence was added for privacy, we designed a solid cedar fence painted a warm beige and added an additional 2' white trellis on the top, the fence was constructed by "Walpole". The pool equipment is concealed by matching cedar fencing and gates in white. We added bronze up lights to the fence by attaching them to the columns for a night time effect, and further lighting around the patio for dining outside.

Travertine natural pavers in a warm beige color. The paving pattern is French Bond that comprises three paver sizes.

The Client loves gardening and using Tropical plants that he winter's over. This vibrant display carries out into the planters displayed along the sides of the pool. More plants were added for year round interest, fragrance and texture.

A J Miller Landscape Architecture was the lighting designers for this project. Electricians installed a comprehensive landscape lighting throughout. Brass cast light fixtures were chosen for their durability and water resistant qualities. Indirect lighting principals were used so that the light fixtures were concealed and blended into the hardscape and plantings.

Designer's Home

Terrain: Flat Ground

The garden is situated in the City of Syracuse, New York (zone 4) along a very old historic street filled with homes built in the early 1900's. Our home is a four square Colonial built in 1911 in the Arts and Crafts period. The garden will be 12 years old this August. There was nothing here before but lawn, overgrown shrubs and dead trees. There was an asphalt pathway that went around the entire outside. We are the owners of this garden. Our family consists of two adults and one 7 years old daughter. We are both landscape architects - garden designers.

The first thing we did was: remove all the lawn and dead trees and shrubs. We planted tall hedges along the sides to contain the space and 900 Boxwoods hedges were planted to form the decorative shaped parterre. Within the two back square of the parterre Holly Cubes are planted that are being shaped and trimmed to form square cubes.

The garden reflects our family as we have introduced the European style of gardens that we are accustomed to while living there. While visitors view our garden as being very different from the gardens locally, we do not find it as curious or unusual, what do find strange is the general lack of structure or "bones" in the design of most gardens we see. The plant materials evaporate in the fall and nothing reappears until late spring. We feel that the garden space should work all year and that plant materials need to earn their space and not be viewed as a two-week delight. Also with the emphases on large sprawling lawns we are looking at showing that a garden need not be reliant on grass, and the upkeep and use of water and fertilizer to sustain this practice.

James Street Garden

Designer: Anthony James Miller & Mariane Wheatley-Miller
Design Company: A J Miller Landscape Architecture PLLC
Location: Syracuse NY, U.S.A.
Site Area: 947.6 m^2
Photographer: Charles Wright

1833

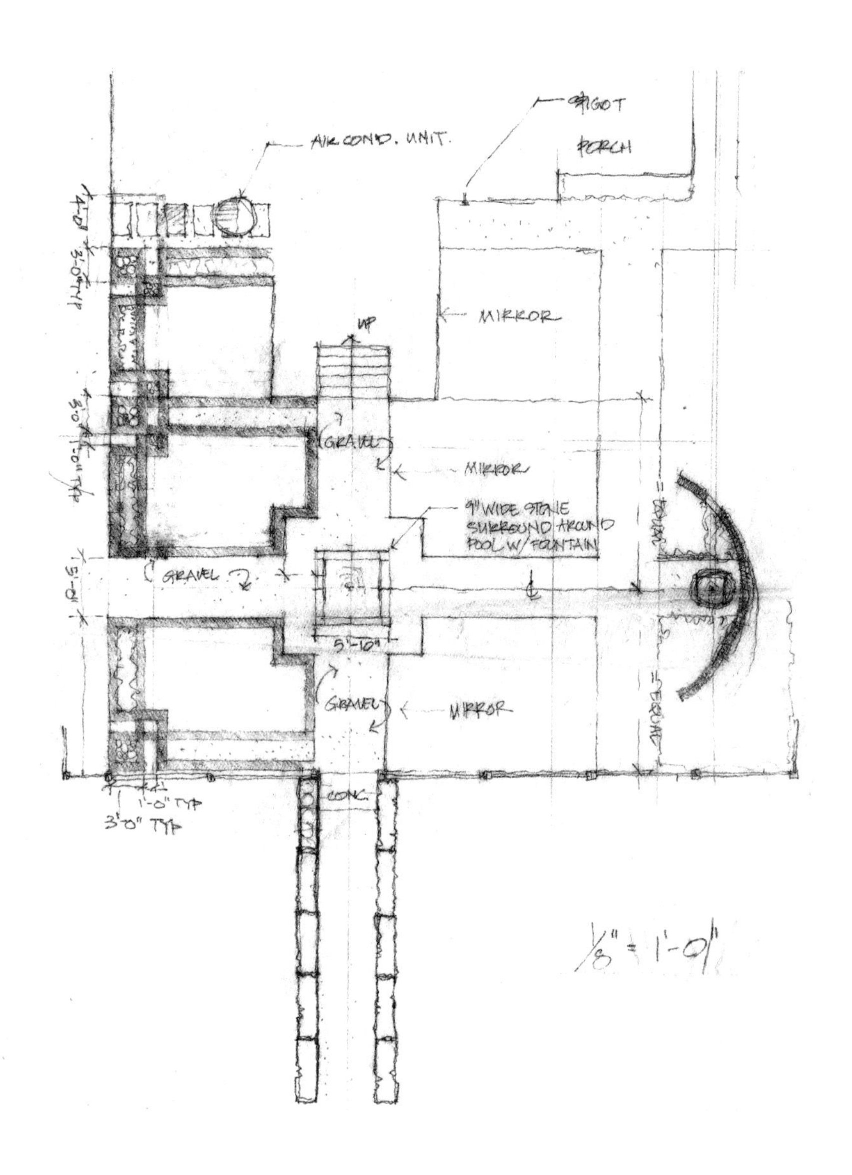

In designing the garden it was important that the design reflected the style and forms of the house. We repeated the 4 square pattern that appears on the stain glass windows into the parterre, also with the viewing areas elevated on the two porches one in the front and the other in the side plus a roof garden. The design of the parterre is viewed and appreciated as you look down upon it. The color palette for the garden is purple, white, yellow, silver, black and green this is to contrast with the house color of yellow with a grey roof.

The style of the front garden is a French formal parterre with seating and a Limestone fountain; the rear garden is contemporary in design; a sunken garden surrounded by a Bamboo grove and shade plants. This area has been designed for outdoor entertainment and cooking with sectional seating arranged around a copper fire pit and barbecue.

1833

The soil is a natural drumlin high on a hill above the street. We like this location as we are in the tree line of the street trees below and have privacy and seclusion from the road. Unfortunately the soil presents its challenges, as it is dry in the summer, stony and poor drainage clay. The existing large trees of Oak and Maple mean we have roots everywhere so digging and planting anything are extremely difficult. We only use organic products on the garden; no pesticides and we have no irrigation so watering is limited.

The Syracuse winters are harsh. We wanted to play with the winter season using evergreen plant materials so we always have forms and structure in the garden.

We have used many climbers by adding a rose garland along one side, we use climbers so we can add another dimension, and this verticality moves the eye up and adds more interest and depth to a garden.

We wanted to incorporate a water feature both to encourage birds and animal life into the garden and also as a white noise for any passing cars.

The limestone center fountain does provide a gentle and soothing sound, we have many birds that bathe and drinking from the fountain and in the summer this water feature is the life of the garden.

The garden is always changing, plants are added and plants are moved around in spring and fall. The rear garden that is north facing and under construction is in a more contemporary style, it contains many shade plants, a clump Bamboo grove, a fire pit and seating area. We also incorporate two steel sculptures and lighting.

Agreeable Little Garden

Terrain: Flat Ground

Located in the residential area of San Siro in Milan, the traditional courtyard of an old and typical "casa di ringhiera" (house with railings), recently restructured to a Bed & Breakfast, has been transformed into a garden available to the guests of the accommodation.

Find the Time

Design Company: INSITU Landscape Design
Location: Milan, Italy
Site Area: 1,200 m^2
Photographer: Luca Baroni

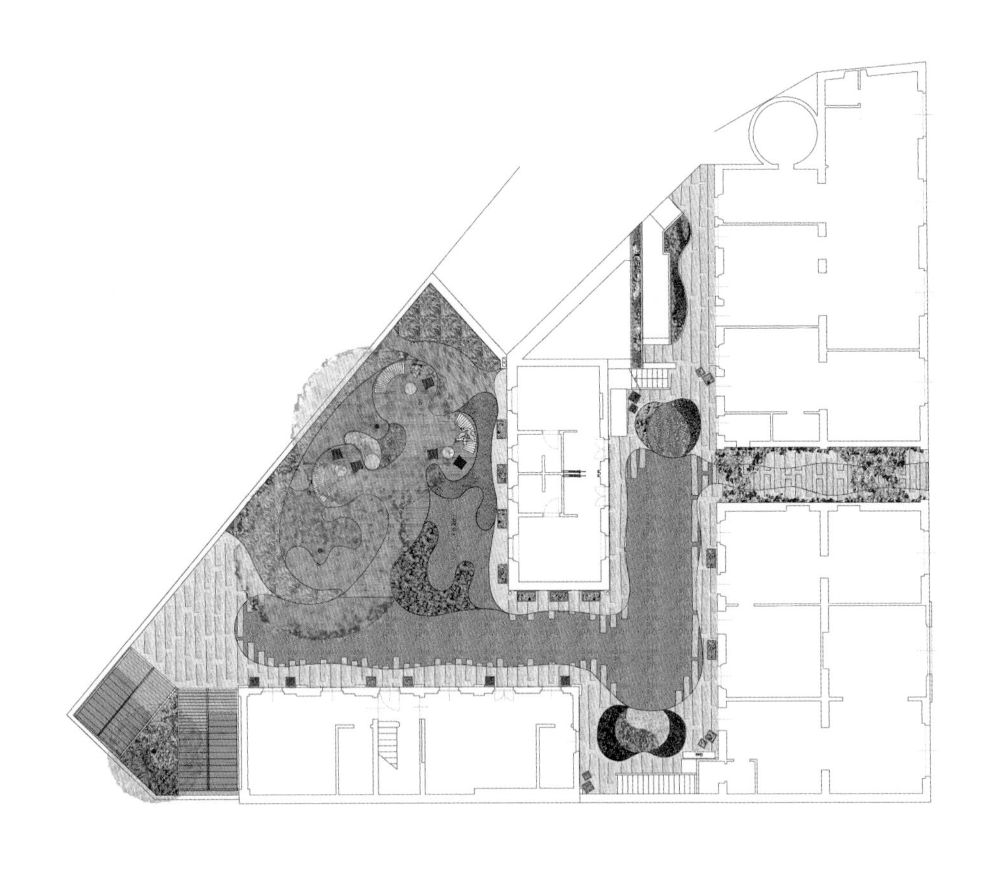

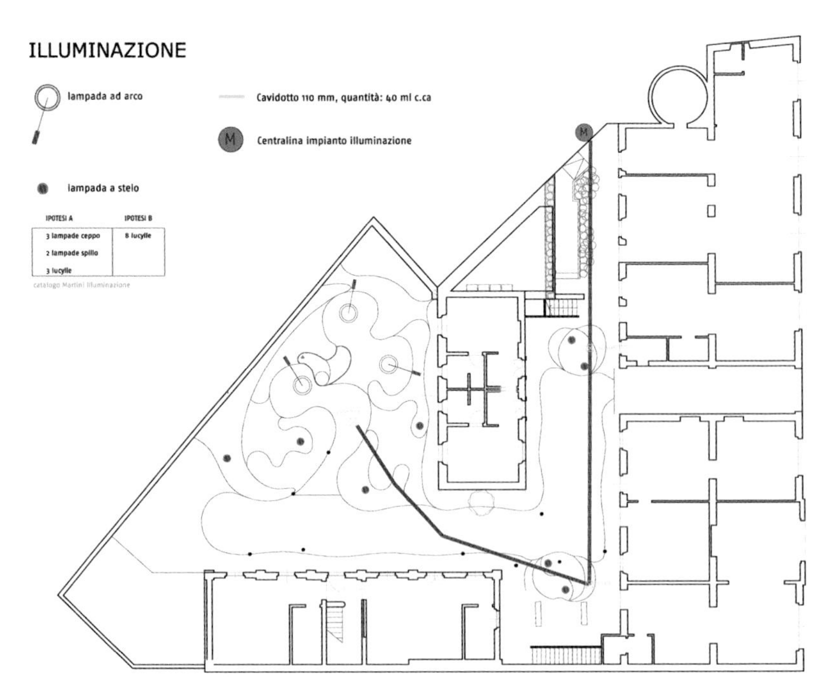
ILLUMINAZIONE
lampada ad arco
Cavidotto 110 mm, quantità: 40 ml c.ca
Centralina impianto illuminazione
lampada a stelo
IPOTESI A
IPOTESI B
3 lampade ceppo
2 lampade spillo
3 lucylle
8 lucylle
catalogo Martini Illuminazione

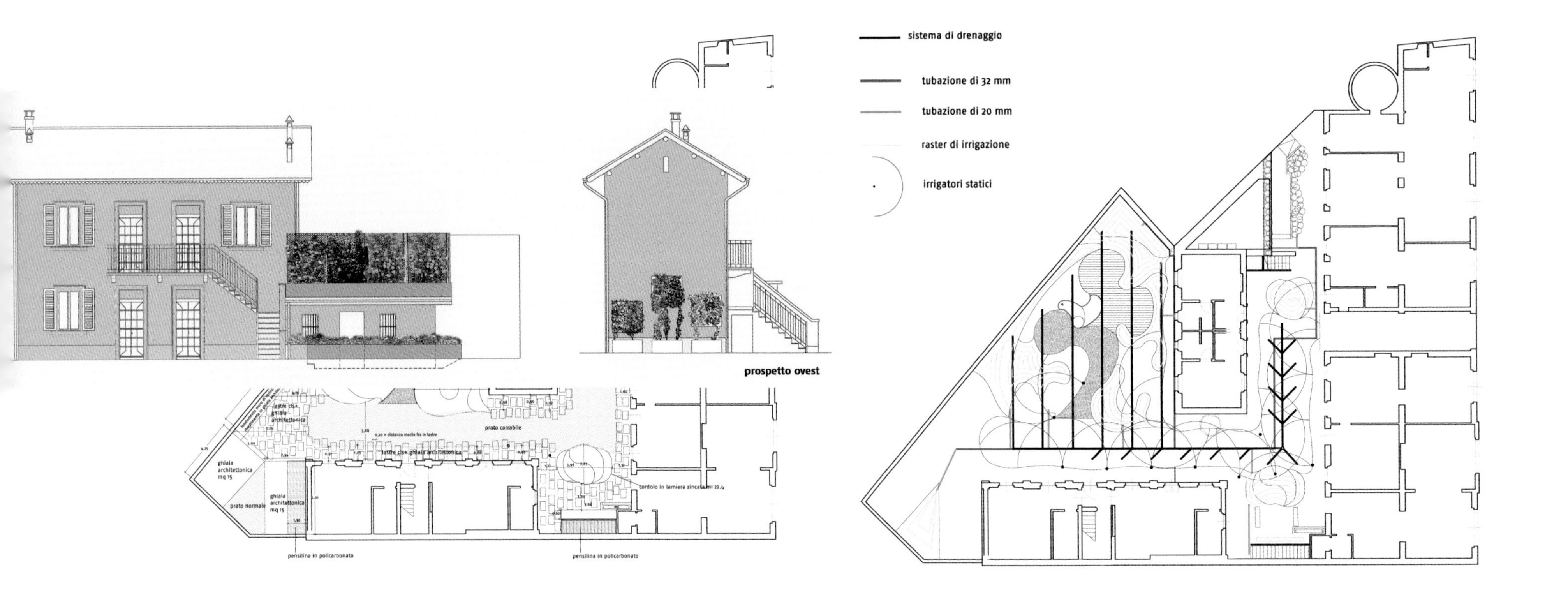
prospetto ovest
lastre cls+ ghiaia architettonica
prato carrabile
0,20 > distanza media fra le lastre
lastre cls+ ghiaia architettonica
ghiaia architettonica mq 15
prato normale
ghiaia architettonica mq 15
cordolo in lamiera zincata ml 22.4
pensilina in policarbonato
pensilina in policarbonato
sistema di drenaggio
tubazione di 32 mm
tubazione di 20 mm
raster di irrigazione
irrigatori statici

Going through a long arcade, from the chaos of city traffic you enter - like magic - this precious green oasis. What was until a few months before a paved yard, used as a parking lot for cars and storage for various materials, has now become a small, enchanting garden.

The floor of the entrance hall was made by placing in an irregular way slabs of concrete on a bed of architectural concrete paving with exposed gravel. From the entrance hall the paving enters the garden continuing along the walls of the houses.

A lawn that can be crossed by the two family cars to reach the two parking spaces assigned to them, connects the two small courtyards that form the garden.

The first courtyard, enclosed by three buildings, can be reached via the paved hall. Two flower beds with flowering plants and aromatic plants are situated at the end of the central lawn. Two metal structures with climbing plants have been added to hide, enhance and value corners of the building structure otherwise uninteresting.

The second courtyard is in the back of the second courtain of buildings and was designed as an area of rest and relaxation; for the morning breakfast and a drink before dinner. Wood and white gravel paved areas are enclosed by patches of bamboo, lavender, Rosa, *Potentilla, Hypericum*. The gentle gurgle of a small fountain cools the hot summer evenings.

MODERN AND NATURAL STYLE

Modern and natural style chases fashion and trend. It focuses on the integration of space layout and natural environment. Its main characteristic is the formal beauty and the simple shape of architectural structures, rather than superfluous decorations. With reasonable space layout and harmonious integration of nature, this style is simple and practical.

Minimalist Art Garden

Terrain: Sloping Fields

Woollahra

Design Company: Secret Gardens of Sydney
Location: Sydney, Australia
Site Area: 600 m^2
Photographer: Nick Watt

Secret Gardens inherited a complicated layout of walls and different levels when first briefed on this project. The overall look was very dated. The existing garden was effectively removed with the exception of a few larger trees. The design focused on a style that was complementary to the architecture of the house at the front entrance. Classic planting with Buxus hedging and Clivia is combined with the use of focal points such as the pot filled with Zoysia that add contrast and interest to the entrance. A more contemporary design was adopted at the back of the property in line with the style of the clients. A young family the design had to meet a number of briefs. Entertaining for the adults and family gatherings, pool for the kids, vegetable and herb garden and of course play areas for the kids.

The design of the pool took into consideration the family size, when designing the size and layout of the pool. The size allows a number of people to enjoy being in the pool together. There is also ample space around the pool to provide access and a place to jump out and sit. A shaded outdoor lounge area provides protection from the sun and a great spot for adults to sit with friends and watch the kids. The pool equipment was designed to be out of sight with a purpose built cupboard adjacent to the pool. This also was designed to store all the cushions that were custom made for the outdoor bench seats and also the clients' outdoor table and seats. This ensures that when not in use they are stored out of sight. A simple thing that is often overlooked in most designs.

The BBQ was designed to be recessed as part of the garden walls and a custom made cupboard and bench incorporated to ensure a seamless finish in the entertaining area. The children are catered for with a cubby house and lawn area to play on. The vegetable and herb garden was kept close to the kitchen. Rosemary was used as a border plant that trails over the edge, and whilst incredibly striking also serves a very functional purpose for the clients. The combination of clean lines in the construction and a focus on the horticultural aspects of the garden has resulted in a timeless design that provides the perfect outdoor areas for this young family to enjoy.

Modern Natural Garden in the Forests

Terrain: Mountainous Region

Private Residence 4

Design Company: Paul Sangha Landscape Architecture
Location: Vancouver, BC, Canada
Site Area: 4,046.9 m^2

This garden design bridges the iconic architecture of Arthur Erickson, who originally designed the home, with the serene woodland of Pacific Spirit Park. The integrity of Erickson's original design was complimented while fresh life was infused into the garden, at the same time celebrating the natural backdrop of dense native forest. The firm wanted the landscape to become a source of inspiration for the clients, while at the same time being sensitive to their practical needs and appropriate to the surroundings. Design strategies such as lush native planting, wildlife habitat restoration, and dramatic opportunities to interact with the surrounding forest encourage an intimate relationship between people and nature. The result is a contemporary garden that feels timeless and rooted in its Pacific Northwest surroundings.

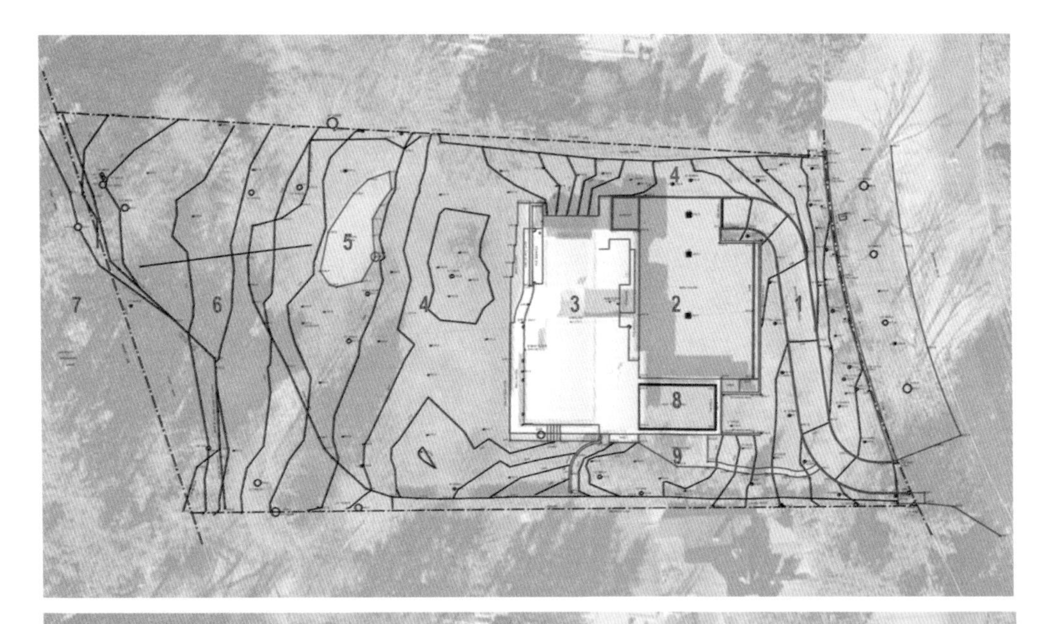

BEFORE DEVELOPMENT

1. DRIVEWAY
2. AUTOCOURT
3. RESIDENCE

4. LAWN
5. GRAVEL PIT
6. RANDOM PLANTING

7. PACIFIC SPIRIT PARK RAVINE
8. GARAGE
9. PLANTED SIDEYARD

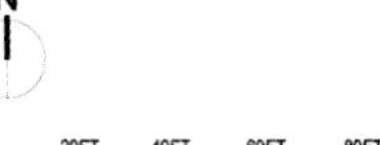

The home was renovated to allow for the needs of the young family that now lives there. During the renovations the clients requested a greater connection from indoor to outdoor spaces. The outdoor space requirements were to provide terraces for entertaining, lounging, and cooking, as well as a large lawn expanse for children to play and a pavilion looking into Pacific Spirit Park.

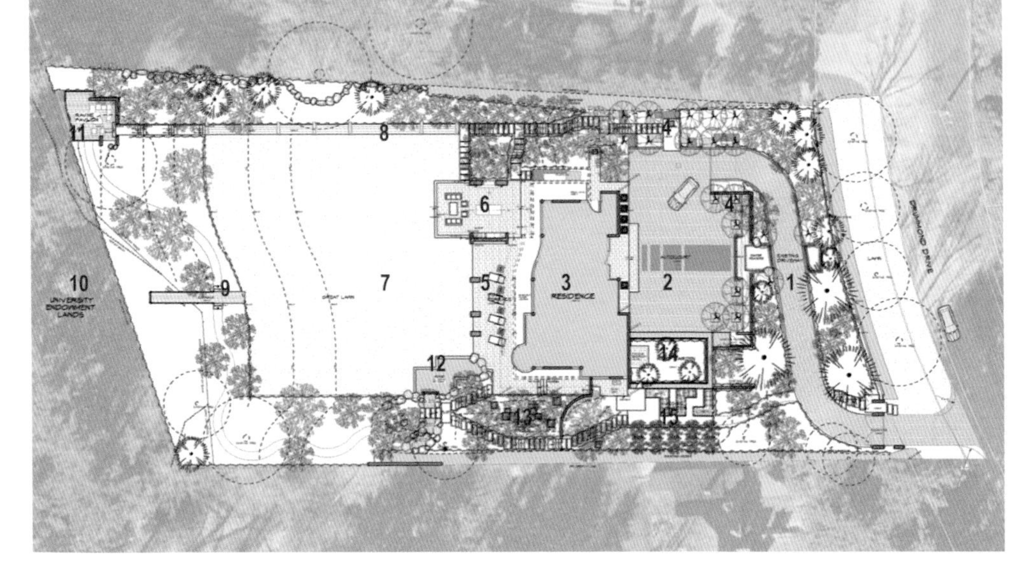

SITE PLAN - AFTER DEVELOPMENT

1. EXISTING DRIVEWAY
2. AUTOCOURT
3. EXISTING RESIDENCE
4. PERSIAN IRONWOOD BOSQUE

5. POOL TERRACE
6. ENTERTAINMENT TERRACE
7. GREAT LAWN
8. ARBOR WALK

9. CANTILEVERED CATWALK
10. PACIFIC SPIRIT PARK RAVINE
11. RAVINE PAVILION
12. POND W/ NATURALIZED WATERFALL

13. CHERRY BLOSSOM KNOLL. RELOCATED EXISTING TREES.
14. ROOF GARDEN OVER EXISTING GARAGE
15. KITCHEN VEGETABLE GARDEN

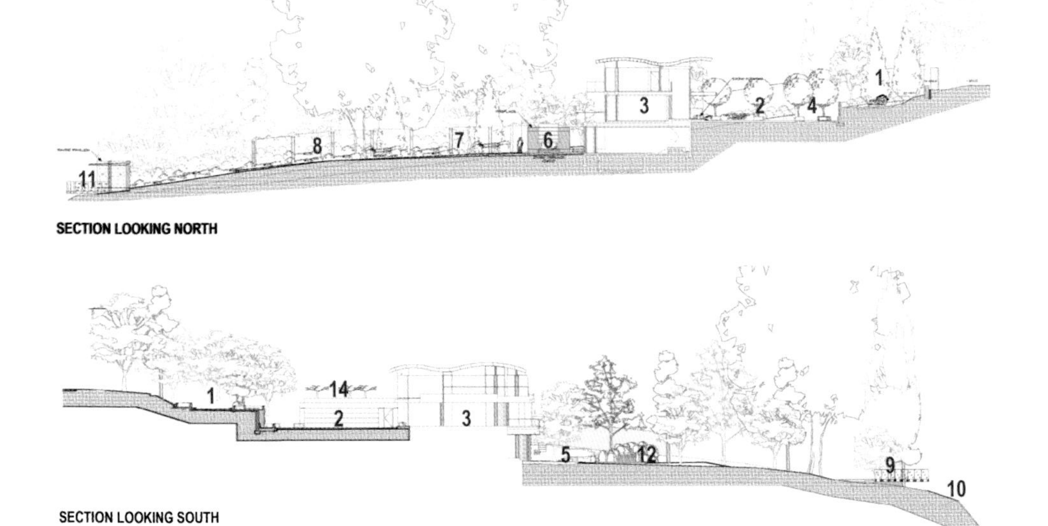

SECTION LOOKING NORTH

SECTION LOOKING SOUTH

SECTION LOOKING SOUTH

1. EXISTING DRIVEWAY
2. AUTOCOURT
3. EXISTING RESIDENCE
4. PERSIAN IRONWOOD BOSQUE

5. POOL TERRACE
6. ENTERTAINMENT TERRACE
7. GREAT LAWN
8. ARBOR WALK

9. CANTILEVERED CATWALK
10. PACIFIC SPIRIT PARK RAVINE
11. RAVINE PAVILION
12. POND W/ NATURALIZED WATERFALL

13. CHERRY BLOSSOM KNOLL. RELOCATED EXISTING TREES.
14. ROOF GARDEN OVER EXISTING GARAGE
15. KITCHEN VEGETABLE GARDEN

A contemporary take on an Art Deco inspired sculptural entry gate greets visitors to the property. At this point, one can already sense the impeccable craftsmanship and detail of the entire project. A high level of customization and care was taken in the design of the address numbers and the mailbox, which complement the entry gate. A clean, minimalist approach to the design is evident as four perfectly aligned stone cubes, which double as fountains, sit at the entrance to the home. The cubes are a slightly playful yet incredibly pure expression of strong geometry. A contemplative and almost still invisible edge water feature spills into, and enlivens, the auto court below. All of the water features are chemical-free in their maintenance. A bosque of Persian Ironwood trees in the auto court unites hard and soft architecture elements.

The hedge forms below explore a cubist expression that expands into the larger landscape. At the south side of the property, a cubist inspired vegetable garden plays on, and reinforces, the geometry of the house. Slab stairs act as sculptural elements, referencing the movement of the entry gate detail, and spill down both side yards to the lawn and terraces in the rear garden. A series of custom designed, stainless steel geometric archways encourage movement through their rhythmic pattern. At night, the arches double as spectacular, LED-lit, sculptural elements, which define the space. The archways also provide space for climbing. The expanse of lawn acts as both open space for the children to play and a foil to both the natural and architectural elements of the site.

A holistic planting approach was implemented. The adjacent ravine of Pacific Spirit Park was infested with invasive plants, which threaten the future of Canadian forests. The firm removed the invasive species, and re-introduced native plants that will strengthen the habitat potential of the area and promote biodiversity of species. No trees were removed during the project, and great care was taken to protect existing trees. A group of mature Cherry trees were relocated from the driveway to the rear yard. Several large trees were introduced, some of which were up to 40 feet tall. The green roof of the garage was replanted using Mountain Hemlock and other native plants.

The striking red of Cornus sanguinea is a contrast to the emeralds of the forest and draws one to a cantilevered catwalk that soars over the ravine plantings below. This extension provides a dynamic and direct relationship between the forest and the home. Imagination is encouraged to grow as one moves down the wood and stainless steel catwalk toward the tree canopy ahead. In contrast, at the northwest corner of the garden, a quiet Ravine Pavilion encourages a quieter experience with nature. The sheltered concrete structure houses a beautiful wood deck with cozy outdoor furniture. The structure was designed to encapsulate the beauty of the forest, and it does just that, cleverly showcasing certain views, while hiding others. The composition of native plants and trees is the focal point and is framed perfectly from inside. Between the catwalk and the pavilion, as the carefully planned views from the rest of the property, this garden's intimacy with its surroundings is an experience like no other.

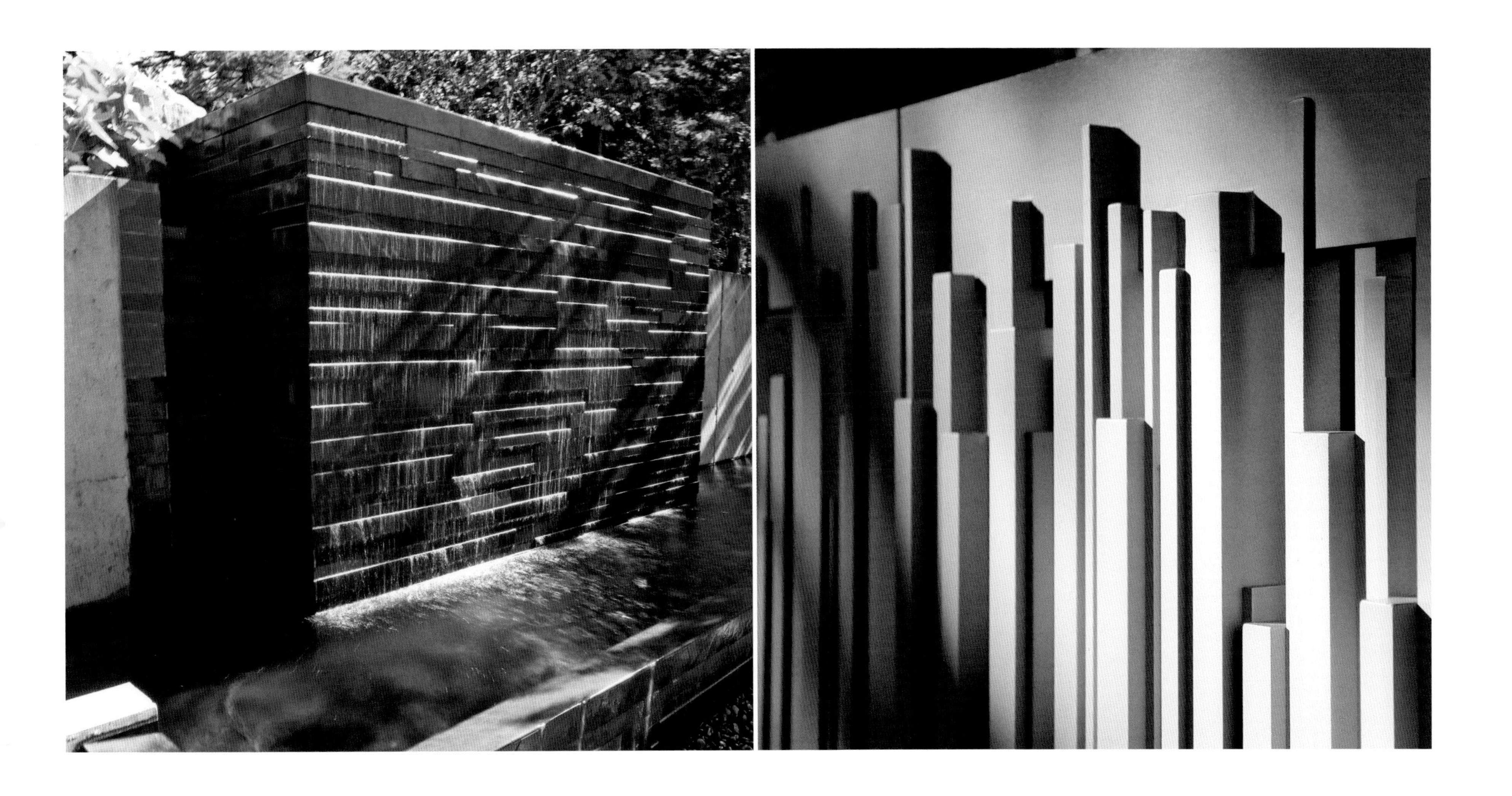

Careful consideration was taken to match the hardscape materials to the home for a seamless transition between architecture and landscape. The fireplace, bar, pavilion and lighting benches are colored concrete that complement the home. The paving material is a rich, natural slate, which also acts as cladding in the water features. Wood and stone details amalgamate the design with the forest, including an impressive permanent stone table in the dining area.

Modern Leisure Garden

Terrain: Flat Ground

Palos Verdes Residence

Design Company: Surface Design Inc.
Location: Los Angeles, CA, U.S.A.
Photographer: Marion Brenner

Designing the garden in the house you were raised in is perhaps the most personal of all projects, and this Southern California landscape, a collaboration between James A. Lord of Surfacedesign Inc. and his mother, is no exception.

Located in Palos Verdes Estates, this residential landscape includes a courtyard, a street landscape, a water feature and a hot tub. The garden reflects the designer's fascination with the sculptural intensity of succulents and unifies the contemporary spaces of the mid-century California ranch house with the Mediterranean ambiance of Palos Verdes Estates.

The courtyard is organized around a stunning California Pepper tree, with sandstone paving that is interrupted by outcrops of bold planting. These "fissures" in the paving are a direct reference to the family's history as the foremost Seismic Structural Engineers in California. The garden is thus a symbolic celebration of both the arid landscape and its tectonic shifts that the family has called home for over forty years.

Beach Landscape

Terrain: Mountainous Region

The 6,475 m^2 Vancouver property is situated on top of a natural bluff overlooking soaring trees, the Pacific Ocean, public beaches and the backdrop of beautiful coastal mountains. It enjoys an infinite view of the Gulf Islands, Sunshine coast and a distant view of downtown Vancouver.

The firm renovated this ocean-side property with the simple intentions of celebrating such pristine surroundings, promoting sustainability and biodiversity, as well as increasing its usability. The resultant garden enhances the full potential of the space and its surroundings. It has become a canvas for artistic expression and a flawless extension of the existing house.

Private Residence 1

Design Company: Paul Sangha Landscape Architecture
Location: Vancouver, BC, Canada
Site Area: 6,475 m^2

The site, prior to the renovation, had most of the functional elements the owners required. However, the previous layout's size and accessibility discouraged the clients from being able to maximize its use. The north banks, comprising one-third of the back yard, were alienated from the rest of the garden by a solid hedge and tennis court fencing. The banks were used primarily to dump years of Christmas trees and garden waste. The owners requested that the garden create smooth flow between indoor and outdoor activities. Being avid art collectors, the owners also wanted their garden to become a gallery of beautiful sculptural pieces for their guests to enjoy. This garden has now become a symbol of their love for art, the environment, and a venue for social interaction and various charitable events.

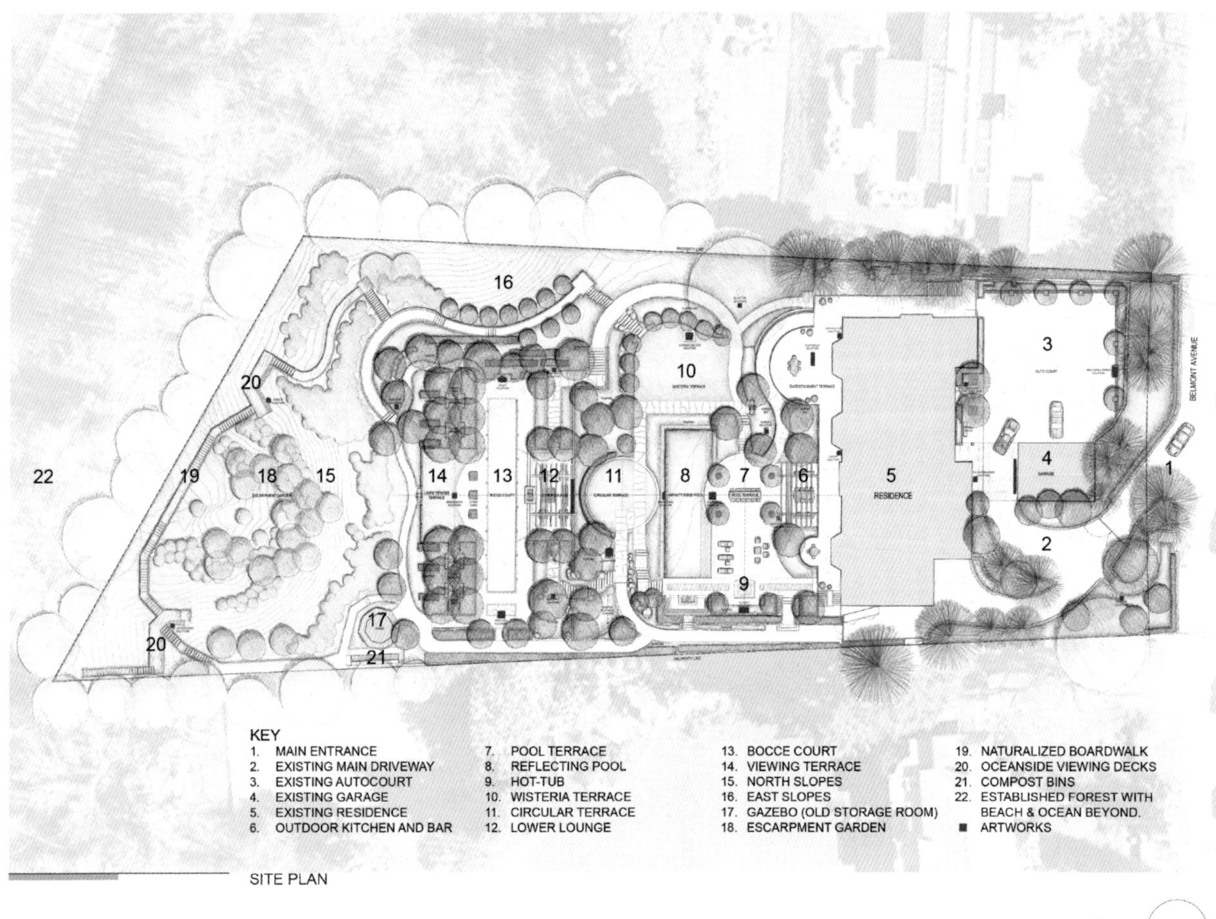

SITE PLAN

The owners committed to retaining existing structures and grades wherever possible. An extensive inventory of existing trees, light fixtures and design elements was made to re-integrate into the new development. Perhaps the single most powerful design gesture was to reshape the pool at its current location to a rectilinear mirror of the sky and landscape beyond. The reflecting pool acts as a beautiful abstraction, amalgamating sky, mountains and the ocean in one canvas within the site. Apart from its aesthetics, the linear pool holds both a functional purpose for fitness swimming, while also creating an architectural backdrop to the outdoor living, kitchen and bar area.

A fine balance is achieved between a modern garden and the wild landscape beyond through a series of interconnected stepping terraces, with one seamlessly blending into another. The northern and eastern steep banks were cleared of all debris and invasive species. After a careful slope and view analysis, a system of boardwalks with strategically located viewing decks was created. This allows the owners, their grand kids, and guests to enjoy the whole property with a varied experience of the terrain, design and planting character. An existing storage structure was modified and re-appropriated to a key garden structure with views to the ocean and mountains.

The planting scheme of the property focuses on providing year round color and texture. A grove of Cornus kousas separates the northern slope with a subtle horizontal line, acting as an echo of the water edge. A substantial planting feature includes the intensive bank stabilization of the north and east banks. Here, select native plantings were chosen for their stabilization properties.

Symphony of Light and Materials

Terrain: Mountainous Region

This unique home was designed by Hotson Bakker Boniface Haden Architects & Urbanistes and is situated on Vancouver's west side. The garden is designed by Paul Sangha Landscape Architecture and the intention behind the garden design is simple: cool green woodland spaces embrace the existing Douglas Fir stand while wide terraces and reflecting pools blur the line between architecture and garden. The open floor planned home uses light as a primary design driver, which is echoed in the garden design. The garden acts as a sanctuary where light and shadow play against texture, achieving a sophisticated, balanced effect. The result is a calm oasis through the intricate weaving of subtle dichotomies.

Private Residence 3

Design Company: Paul Sangha Landscape Architecture
Location: Vancouver, BC, Canada
Site Area: 1,214 m^2

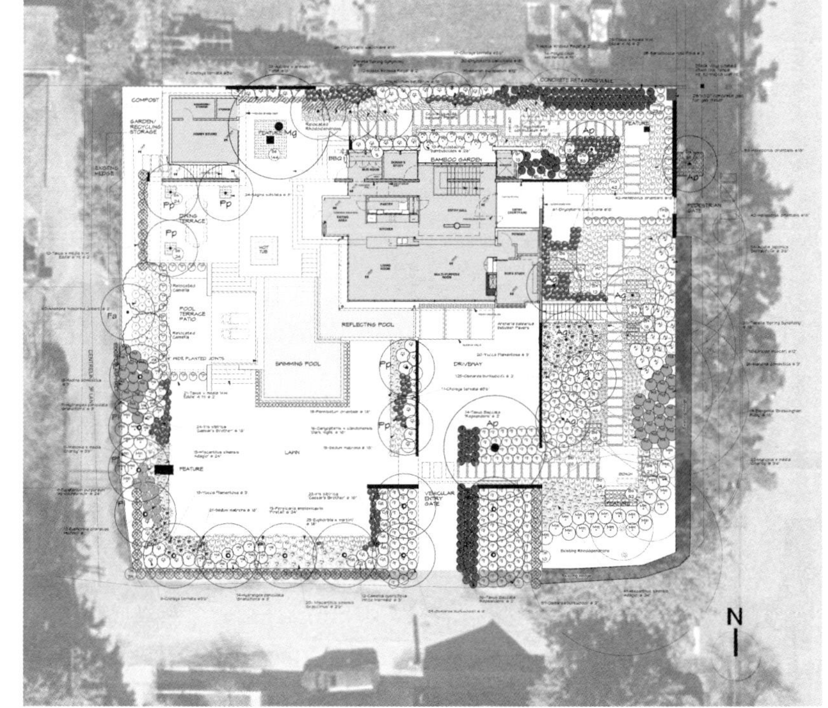
COMPOST
GARDEN/
RECYCLING
STORAGE
BAMBOO GARDEN
BBQ
DINING
TERRACE
POOL
TERRACE
PATIO
WIDE PLANTED JOINTS
REFLECTING POOL
SWIMMING POOL
DRIVEWAY
LAWN
FEATURE
VEHICULAR
ENTRY
GATE
CONCRETE RETAINING WALL
Relocated Camellia
Relocated Rhododendrons
Existing Rhododendrons
N

The entry is layered from the street and the frosted glass and steel entry gate play on the lines of the architecture, while allowing light to penetrate. Geometric paving stones reinforce the vertical strength of the trees, while open planted joints allow rainwater to infiltrate the soil. Concrete pavers are embedded into a thick planting of rich moss, lending softness to the hardscape. A unique series of natural and urban courtyards ultimately connect to terraces, which spill down to the lawn. Clean lines and pure forms characterized the concrete work, which seems to extend from the home outwards. Subtle, yet effective details make a significant impact. A crisp strip of grass cuts through the concrete terrace, framing and softening the space.

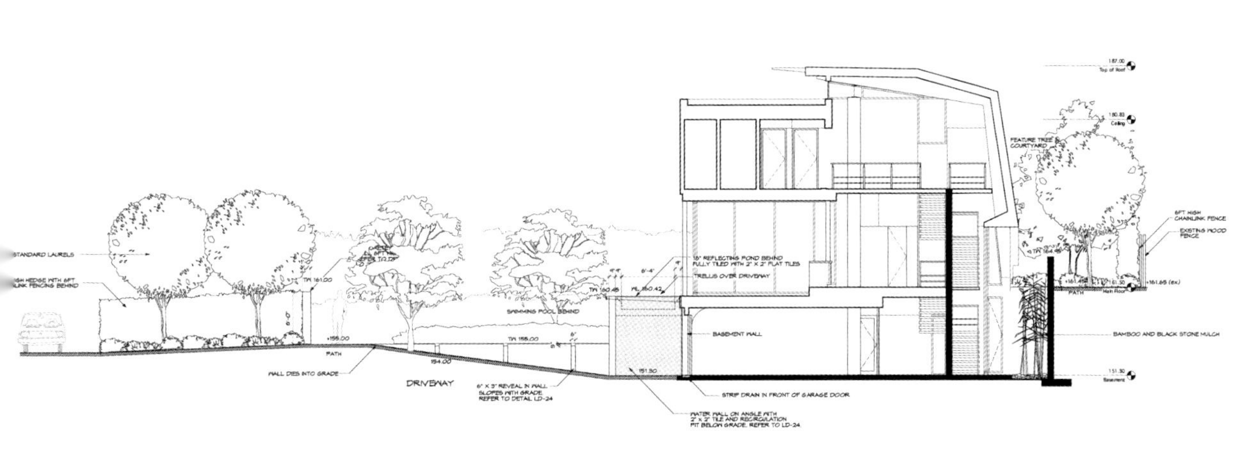

A thick, textural burst of planting frames the lawn. From the lawn, the reflecting pool acts as a dramatic contrast to the concrete of the house and the terraces. Sky, trees, and planting are framed in the reflection of this geometric form. The multi-layered water feature acts as a reflective pool from the view inside the home, and as a sculptural piece in its own right from the lawn. Deep grey tiles lend weight to this water feature, as it visually appears deeper than it is. The concrete stairs and terraces seem to sink into the feature, amalgamating architecture and garden.

Throughout the garden, sculptural pieces have been integrated into this balance of light, texture and weight. The result is a fresh, inspirational and ultimately rejuvenating outdoor experience.

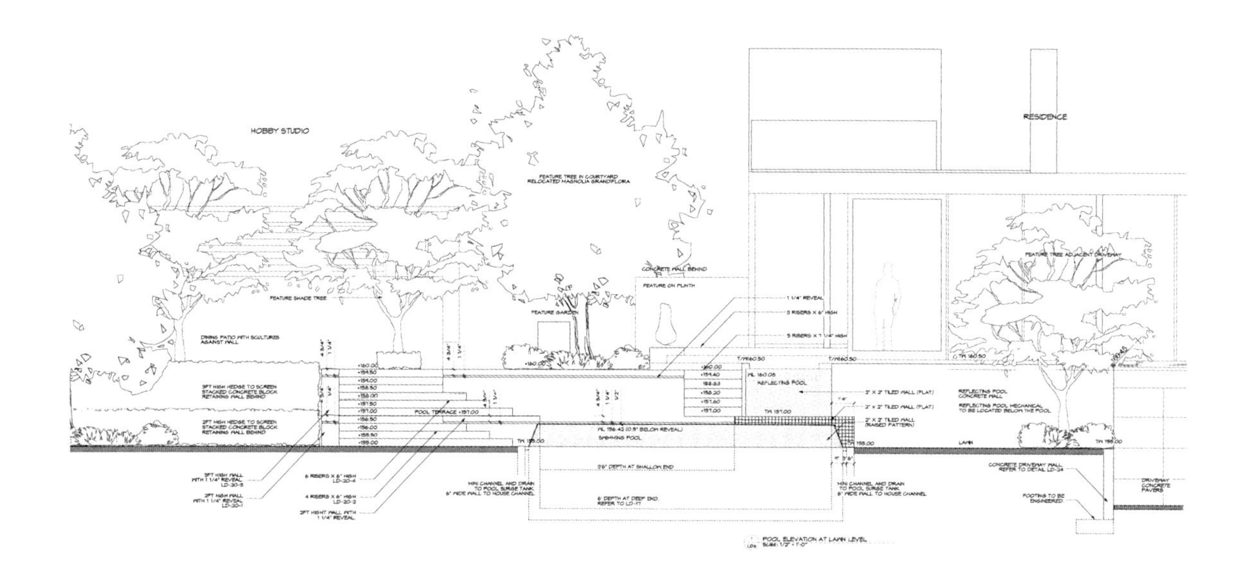
HOBBY STUDIO
RESIDENCE
FEATURE TREE IN COURTYARD
RELOCATED MAGNOLIA GRANDIFLORA
FEATURE SHADE TREE
DINING PATIO WITH SCULTURES
AGAINST WALL
CONCRETE WALL BEHIND
FEATURE ON PLINTH
FEATURE GARDEN
FEATURE TREE ADJACENT DRIVEWAY
3FT HIGH HEDGE TO SCREEN
STACKED CONCRETE BLOCK
RETAINING WALL BEHIND
2FT HIGH HEDGE TO SCREEN
STACKED CONCRETE BLOCK
RETAINING WALL BEHIND
POOL TERRACE +157.00
WL 156.42 (0.5" BELOW REVEAL)
SWIMMING POOL
3'6" DEPTH AT SHALLOW END
6' DEPTH AT DEEP END.
REFER TO LD-17
3FT HIGH WALL
WITH 1 1/4" REVEAL
LD-20-3
2FT HIGH WALL
WITH 1 1/4" REVEAL
LD-20-1
6 RISERS X 6" HIGH
LD-20-4
4 RISERS X 6" HIGH
LD-20-2
2FT HIGHT WALL WITH
1 1/4" REVEAL
MINI CHANNEL AND DRAIN
TO POOL SURGE TANK
6" WIDE WALL TO HOUSE CHANNEL
1 1/4" REVEAL
3 RISERS X 6" HIGH
5 RISERS X 7 1/4" HIGH
WL 160.08
REFLECTING POOL
2" X 2" TILED WALL (FLAT)
2" X 2" TILED WALL (FLAT)
2" X 2" TILED WALL
(RAISED PATTERN)
REFLECTING POOL
CONCRETE WALL
REFLECTING POOL MECHANICAL
TO BE LOCATED BELOW THE POOL
LAWN
CONCRETE DRIVEWAY WALL
REFER TO DETAIL LD-24
FOOTING TO BE
ENGINEERED.
DRIVEWAY
CONCRETE
PAVERS
POOL ELEVATION AT LAWN LEVEL
Scale: 1/2" = 1'-0"

Residence Greenwich CT

Design Company: Doyle Herman Design Associates
Location: Greenwich, Connecticut, U.S.A.

A Unique Fusion of Architecture and Landscape

Terrain: Mountainous Region

The contemporary house, built in 1998, is a standout in the context of its surrounding traditional neighbors. The owners of this Connecticut home, worked with Kaehler/Moore/ Architects (K/M/A) for both the initial construction in 1998 and later a renovation in 2006. A stylish new landscape was created to compliment the renovation by DHDA. The intent of the addition/alteration and the landscape was to build upon and clarify the relationships between interior and exterior habitable space in a suburban context. The addition set up the building's relationship with its immediate surroundings by occupying the border between two green "rooms"; it not only added to, but also redistributed program; and it created an entry portal which relocated one's moment of arrival to the entry courtyard.

DHDA redesigned the landscape with a unifying plan and it was important that the new designed spaces would belong to both the architecture and the landscape. To the side of the house, an aerial hedge of Pyrus calleryana bordering the driveway broke up the building mass and added a horizontal emphasis that was carried throughout the design. The cedar lattice fence, a critical feature of the overall plan, served to demarcate the property line. This custom fence was utilized to establish a degree of enclosure in all of the landscape spaces. Repetition of materials and patterns ensured that the outdoor spaces flowed from one to another seamlessly.

The project's building masses, or program "solids", were connected by a series of ribbon windows that denote a transitional "void" space and became the design's spatial thread. It is not the program "solids" themselves, but rather the paths between them - the transitional "voids" - that activated the site and the program. This produced physical and visual connections that, in turn, altered both. The addition encouraged a dynamic and productive dialogue that intertwined landscape and building(s) with interior movement and activities.

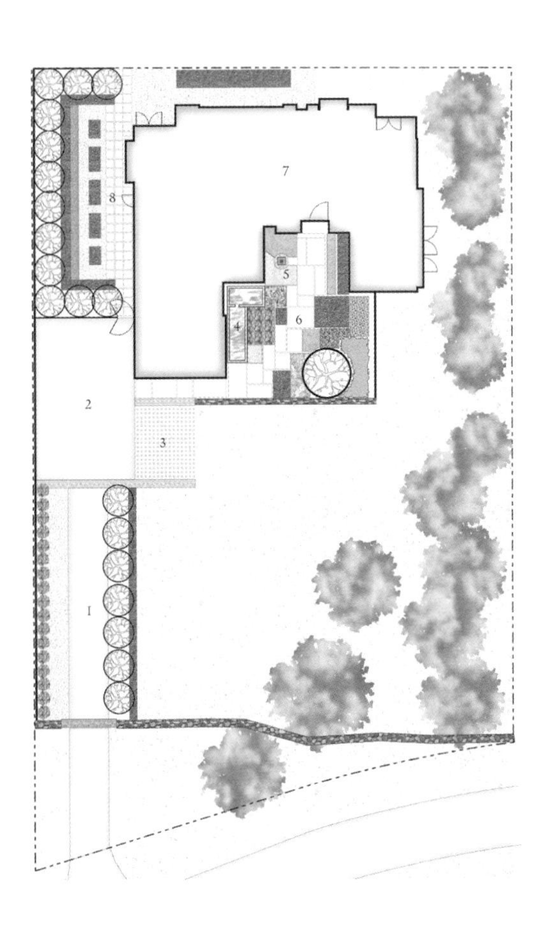

The entry courtyard was defined and enclosed by both the custom fence and a low masonry wall. Two types of stones were used to construct this wall and the cantilevered wall coming off of the house; a rough faced blue stone and a precisely sawn bluestone. The horizontal stone pattern was then repeated as an inlay detail in the driveway. Detailing in the entry courtyard space also included a water feature element. The subtle noise and the movement of water from one trough to another added to the restfulness of the space. Careful considerations were given to appropriate plant combinations. These included clipped groupings of boxwoods, yew and barberry all offset by tufted, blue fescue grasses, and textural perennial plantings of Stachys byzantina, Carex morowii and Sanguisorba menziesii. The main asymmetrical path of oversized bluestone pavers made one feel like they were walking through a Mondrian painting towards the front door. Cream colored gravel was chosen to create a separation between the differing materials but again was part of the thread tying these spaces together.

LEGEND
1. Driveway
2. Parking Courtyard
3. Parking
4. Water Troughs
5. Sculpture
6. Arrival Courtyard
7. Residence
8. Private Courtyard

A Switchback & Metal Sustainable Garden

Terrain: Sloping Fields

Liberty Hill Residence

Design Company: Surface Design Inc.
Location: San Francisco, California, U.S.A.
Photographer: Marion Brenner

Marked by its dramatic sloping topography, this private garden in San Francisco's Liberty Hill has dedicated areas for entertaining and children's play, which are defined and navigated by an innovative arrangement of hardscape materials. Cor-Ten steel boxes serve as retaining structures and planters, extending along the site's perimeter and penetrating the surrounding wood fence. Taken as a whole, the back yard is an abstracted allusion to the hilly city itself, which appears to fall away behind the garden when viewed from the upper terrace or out the back windows of the house.

The Cor-Ten boxes and concrete walls were added to existing stone retaining walls to create a series of switchbacks that guide the journey from the house into the gardens. Erected mid-century, the stone walls form a natural part of the vocabulary of rugged and refined materials, stone, concrete, wood, and steel, while uniting the site's past and present incarnations in a visually harmonious way. The steel boxes, which harbor plant life up top, descend into the ground plane and insert into the redwood fences, a geometric composition enhanced by the slanting shadows that are cast through the variegated slats. The dimensions of the slats are echoed in the board form pattern on the concrete walls that wrap into the fence.

Seeking to maximize permeability for the overall health of the garden, the large children's area is lawn and the entertainment terrace is decomposed granite. The planter boxes not only afford the homeowners an opportunity to do some hands-on gardening, they also attract pollinators such as birds and butterflies. The concrete steps are framed by runnels to enhance drainage (while highlighting the sloped nature of the garden), and a native sedge called Juncus helps purify run-off before it enters the groundwater system. Because the steel boxes are framed by the runnels, they appear almost to rise from below the ground plane, adding depth to the composition.

As the homeowners spend as much time looking down on the garden from their home as being in it, they also appreciate its sculptural quality, which is punctuated by plant material such as Japanese maples and the softening presence of shade-lovers such as ferns, irises, viburnum, and anemones.

The Lakeside Residence

Design Company: AguaFina Gardens International
Location: Suburbs of Detroit Michigan U.S.A.
Photographer: George Dzahristos

Stone and Steel Garden

Terrain: Sloping Fields

Aside from the aesthetic nature of the garden, there were numerous engineering challenges. The house was constructed on unstable former marsh land. In order to stabilize the area, the entire garden space was excavated down to a depth of approximately 15-20 feet to remove organic material and then pylons had to be installed to support the steel walls and other structures.

This garden is a blend of modern design and organic elements. The overall design concept for the lakeside garden was developed by taking a cue from an architectural detail on the house of a contrasting brick that forms a half radius. This motif was then transferred to the landscape by creating a series of interpretive concentric partial radii using corten steel. The idea initially came from questioning of what would the space look like if that radius fell over onto the ground...how would the shape be interpreted.

The walls formed by the steel occasionally intersect several large stone boulders, a combination of Hematite and Iron Ore, which were harvested from the Upper Peninsula of Michigan. The planting spaces are filled with masses of perennials and mature specimen ornamental trees.

The theme of stone and steel continues throughout the garden and is accented by the 3 water features and numerous sculptures placed on the site.

Mountain Villa of Modern and Natural Style

Terrain: Sloping Fields

Rieder's Residence

Design Company: Surface Design Inc.
Location: San Francisco, California, U.S.A.
Photographer: Marion Brenner

The designers transformed a steeply sloping - and seldom used - back yard into a functional outdoor space that offers three distinct - and distinctive - experiences of the garden. The designers approach offers a poetic solution to a problem endemic to many residential gardens in San Francisco: namely, how to create usable space on a steep, underutilized site - exploiting rather than compromising its unique, intrinsic character.

A couple with two young children had lived in their house for some five years before engaging the designers to transform their steeply sloping.

The hill embarks fifteen feet from the house, which feeds into the garden on two levels. Entering through the garage from the street, the first view is of a retaining wall, which was in dire need of replacement. On the main living level, the master bedroom opens onto the primary expanse of usable flat space. The challenge - and opportunity - was to create a cohesive plan for the vertical space, creating an experience of the garden below, providing a hospitable gathering space adjacent to the house, and providing access the top of the site, with its postcard views of the ocean.

On the lower level, replacing the crumbling retaining wall offered an opportunity to incorporate the water feature requested by the homeowners. A fountain is embedded into two tiered walls; as water passes through bronze weirs it cascades down into pools, creating a pleasant aural backdrop throughout the garden. Visually integrated into the natural environment, the board-form concrete used for the walls has a wood-grain pattern that lends textural integrity, providing erosion control without the weight and monotony of a monolithic slab. At its base, the wall is massive enough to plant in, replacing a once desolate space with a kind of tranquil green grotto and contemplative place to pause.

Steps lead up to the main level, an expansive, semi-sheltered outdoor room ideal for dining and a safe play area for the children. Here the goal was to retain an existing oak tree and create harmonious plantings reflective of a natural Northern California plant palette, with native grasses and other drought-tolerant plants surrounding the oak with a sea of muted greens and dusky plums that harmonize against the backdrop of stained black cedar decking and walls.

From here, the designers continue the circuit that embarked in the grotto space below, with stairs gracefully leading up to a viewing platform. Previously inaccessible due to the steep grading, this panoramic aerie offers sweeping views of sunsets over the Pacific Ocean and landmarks such as the Golden Gate Bridge, providing a sense of place within the neighborhood, the city, and beyond.

Planting themes in the garden vary by location. The quiet lower grotto is composed of geometric concrete paving, metal planters, and lush, shade-loving plants, such as *Woodwardia radicans*, *Helleborus orientalis*, and *Persicaria* "Red Dragon". The sunny hillside is covered with a crimson sea of Tradescantia pallida accented with a mix of *agaves* (*Agave attenuate*, *Agave parryi*, and *Agave* "Blue Glow"); Aeoniums and Sedums. The edges of the dramatic hillside and board-formed walls are softened with wispy plumes of *Pennisetum alopecuroides*. A small path of square concrete pavers leading to the top deck appear to float above a soft bed of naïve fescues and are bordered by a mix of Deer Grass (*Muhlenbergia rigens*) and Kangaroo Paw (*Angiozanthus flavidus*).

Simple Seaside Garden

Terrain: Seaside

Private Residence 5

Design Company: Paul Sangha Landscape Architecture
Location: Vancouver, BC, Canada
Site Area: 743 m^2

This urban garden in Vancouver's west side is situated on a pebbly beach looking to the spectacular North Shore mountains and downtown Vancouver. The garden is about duality, effectively balancing its function as a platform for active city life with its immersion in its majestic natural surroundings.

Both the landscape and residence were new construction, which allowed an integrated approach to house and garden. The property slopes to the beach and views, resulting in a garden that cascades into the natural environment. The desire for balance and serenity was foremost in developing the garden. Simplicity in geometry and planting were the key in allowing garden and view to coexist. Water became the binding agent in the garden, bringing reflections of the sky, plants and greater landscape into the property.

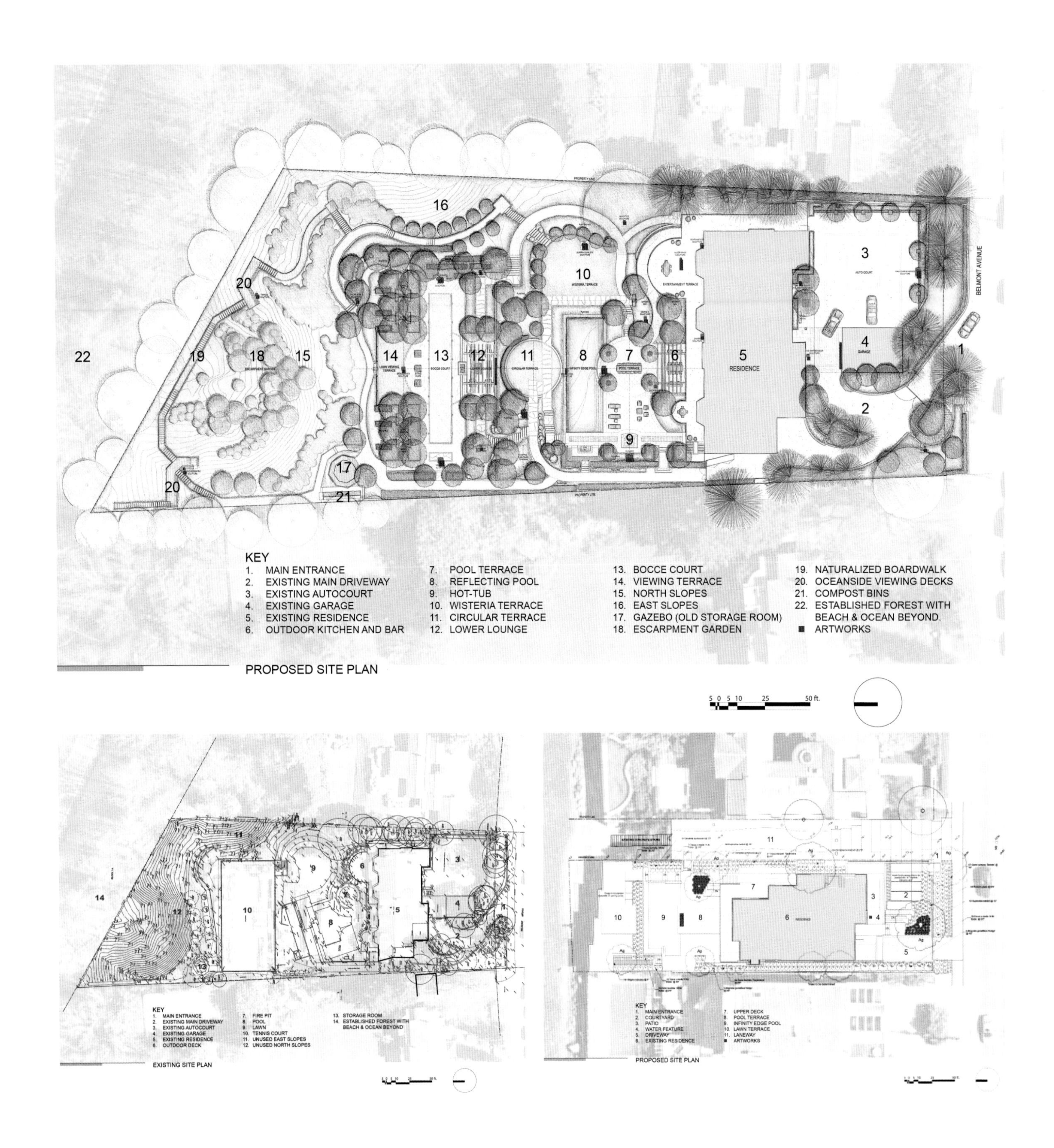

Yew hedges and frosted glass screens hide the entry courtyard creating silhouettes of trees and forms that intrigue onlookers with what may be behind. A shimmering water wall welcomes guests into the courtyard while cascading limestone stairs draw them to the front door. Planting in the courtyard is simple and strong, providing organic balance to the crisp lines of concrete and stone. Anchoring the courtyard are large cinnamon colored paperbark maples that reference the native arbutus.

This integrated landscape design accomplishes a sense of infinite space in an urban setting. The beach garden is a layered tapestry of concrete terraces, a reflecting pool and rich green grass moving across the field of vision. These strong horizontal lines anchor the magnificent vista beyond. A bronze sculpture reinforces the horizontal lines while providing foreground to the composition. Red weeping Japanese maples frame the view and their mounded forms gently echo the mountains beyond. Indirect circulation through the garden encourages guests to engage with the ever-changing vignettes of garden, view and architecture.

Wittock Residence

Design Company: ASPECT Studios & Tzannes Associates
Location: Point Piper, Sydney, NSW, Australia
Site Area: 2,000 m^2
Photographer: Simon Wood

A "Sauvage" Garden

Terrain: Seaside

The unique brief was that the spectacular garden overlooking Sydney Harbour should embody the notion of the "Sauvage" (which is French for savage or wild). The landscape needed to be a range of things including informal, relaxed, bold, eclectic and expressive of its genius loci.

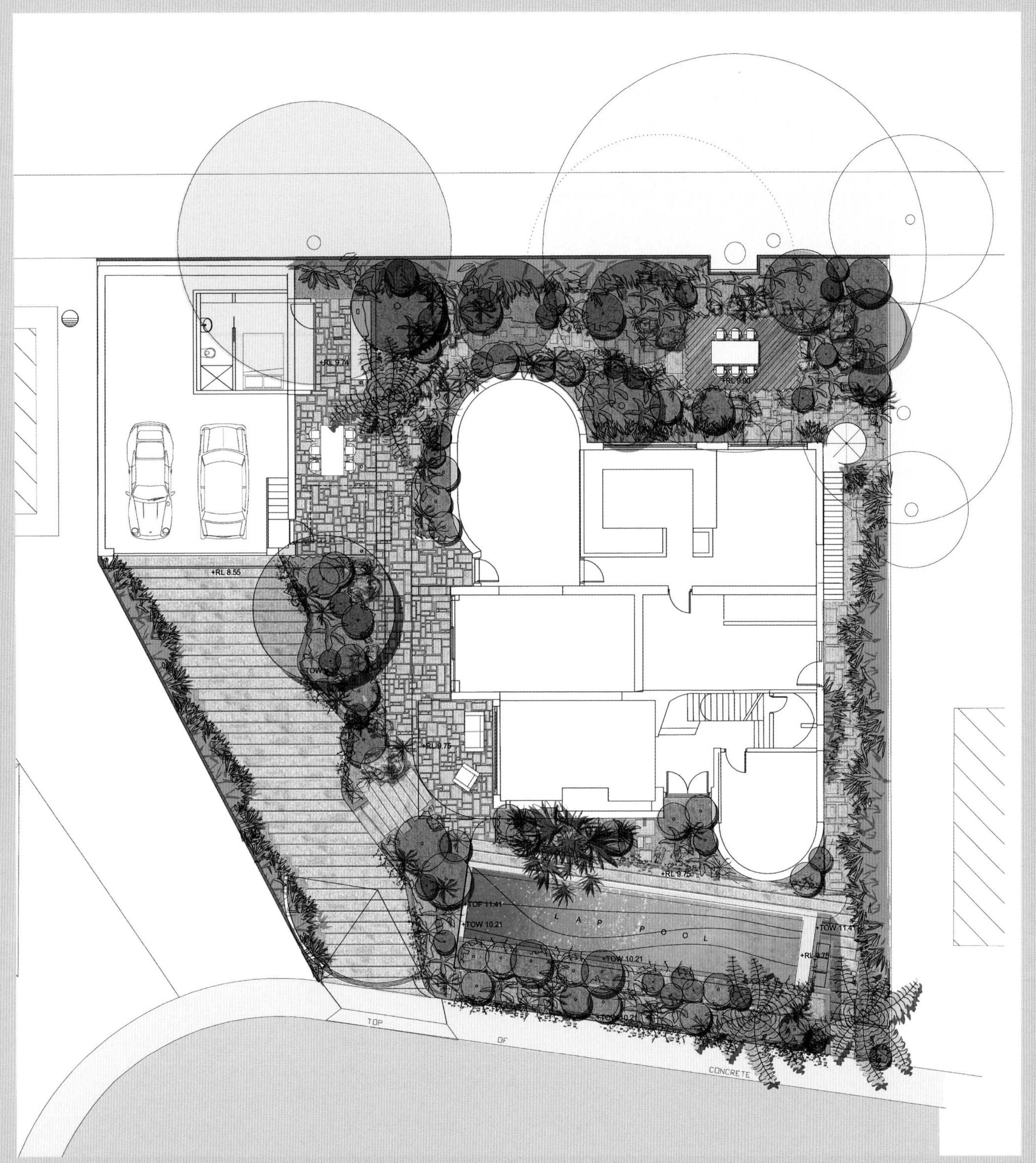

To achieve the notion of the "Sauvage", we installed planting at densities which result in vigorous and lush coverage, particularly in the sheltered areas. The design creates a sense of connected garden rooms on this multilevel site.

A stacked blue-grey stone wall defines the eastern edge of the driveway and wraps around to the front of the property. Thick slabs of basalt form the main staircase from the driveway up to the entry terrace. An in ground timber sleeper deck in front of the rear entry door of the house provides an outdoor dining area amongst the greenery.

The gardens work perfectly in sympathy with the architecture enhancing the distinctive curve of the house and the large terrace overlooking Sydney Harbour. The careful plant selection suits site conditions and the detailing resolve level changes, creating consistency and flow around the house.

Centennial Park

Design Company: Secret Gardens of Sydney
Location: Sydney, Australia
Site Area: 750 m^2
Photographer: Adriano Carrideo

This property is located in a historical part of Sydney around Centennial Park. Entrance to the back yard is either via the house or through a side passage. The sculptural gate at this entrance is the introduction to a new style in the back part of the property. The back garden style is contemporary to suit this modern, young family.

The Fashion Garden

Terrain: Flat Ground

The design focused on ensuring the front garden was complementary to the period of the home with simple, formal plantings combined with clean lines. A garden bed adjacent to the pool adds a horticultural element into a structural area and softens the overall look. This also carries through to the entertaining area and down the entire wall linking areas through the use of plants. The clean lines of the formal and classic planting provide a timeless design. And with all the different areas to entertain, or laze about, there is somewhere to suit every mood.

The design incorporates the ultimate outdoor entertaining area complete with large swimming pool, outdoor lounge area with outdoor TV and fireplace so this can enjoyed all year around. A separate BBQ and dining area was designed with custom built cabinetry. The oversized, comfortable lounges allow adults to sit and relax whilst having full visibility of the kids in the pool. Overhead is a retractable awning, ensuring that in the hot Sydney summer a shaded spot is provided. The pool adds interest and an element of play for the kids with stepping stones that take you to the between the pool and the spa.

Blend of Water, Fire, Rock

Challenging for this site was creating a small intimate gathering area within the context of a larger setting. Set on a flat piece of land adjoining a very popular lake, and at the same general level as the lake, the back of the home was quite exposed.

Terrain: Flat Ground

Modern Garden

Design Company: AguaFina Gardens International
Location: Suburbs of Detroit Michigan U.S.A.
Photographer: George Dzahristos

The challenge was to design a landscape that would accentuate the unique architecture of the home, yet not be overshadowed by the strength of the building. By using very simple but strong forms, the landscape accentuates the architecture and the house stands as a compliment to the garden, fusing two distinct elements into one form.

The earth was shaped and contoured in order to balance out the curves found throughout the building design. By providing clear examples of positive and negative space, sculpted pines were used for an additional layer of architectural interest. Simple placement of stone adds an organic element to the garden.

The custom designed planters are sheathed in lead-coated copper, borrowing from an architectural detail found elsewhere in the building. Their geometric shapes mimic the glass block details of the home and add direction to the existing walkway and entrance.

The obvious choice for lighting in this modern garden would have been to directly up-light all the individual specimen trees and key landscape features. Instead, the house was chosen as a canvas in order to accentuate the features of both the landscape and the architecture. The silhouette of a single sculpted pine is contrasted against the backdrop of the house and provides a simple cultural element. By using light to focus on the interplay between the building and the plants, the synergy between the unique architecture of the home and the landscape is established and the two disparate elements are fused together into a single visually compelling statement.

In order to achieve an intimate and nestled feel, the area for this gathering space was "sunk" by slightly lowering the grade and creating natural berming with mass plantings. Further accentuating this nestled feeling, large monolithic stone basalt pillars amplify the lower aspect of the space. Polished on one side and natural on others, the polished surface reflects both fire and the natural sun as it sets. These pillars are set upon larger base pieces, also of basalt, that are incorporated within the retaining wall designed as a seating area. Other uses of stone integral to this project includes contrasting black Asian gravel along with finer texture and lighter colored pea gravel. Hand selected weathered limestone boulders accent and anchor the garden and are also used as stepping stones through the mass planting of fountain grass and other perennials.

This modern garden fuses three strong elements - Water, Fire and Stone. Central to this garden is the fire and water feature emanating from a large basalt stone that has been cut, hollowed and polished. Water flows over the sides of the piece into the basin below which encompasses the entire pea gravel walkable surface that is used for the patio area. This surface acts as collection zone to recharge and replenish the fountain with rainwater while gravel functions as part of the filtration. Unlike other fire and water features where the fire and water are separate, in this system the water and fire are integral. The water is infused by gas and then the gas "off-burnt", using a proprietary technique that AguaFina developed for this effect and has used on similar projects.

Diverse Modern Garden

Terrain: Mountainous Region

Palo Alto Residence

Designer/Design Company: Steven Ehrlich (architect), CMG Landscape Architecture
Location: Palo Alto, California, U.S.A.

A new residence situated on a corner residential lot demanded a landscape that performed in a variety of ways.

The architecture established four distinct exterior environments. The landscape design reinforces this site structure by establishing places of distinct character, sometimes ordered by the geometry of the building and sometimes playing against it. The entry court was developed around the idea that the sculpted space would be compelling and something to experience at any time. The space is typically empty, but it is the first thing you see arriving at the house. On occasion the court must accommodate up to 100 people for black-tie parties. An undulating and pleached carpinus hedge frames and defines the space(s). Over time the hedge will have more presence, not only as it gets bigger, but also as each branch grows into and fuses with its neighbor to create one living object.

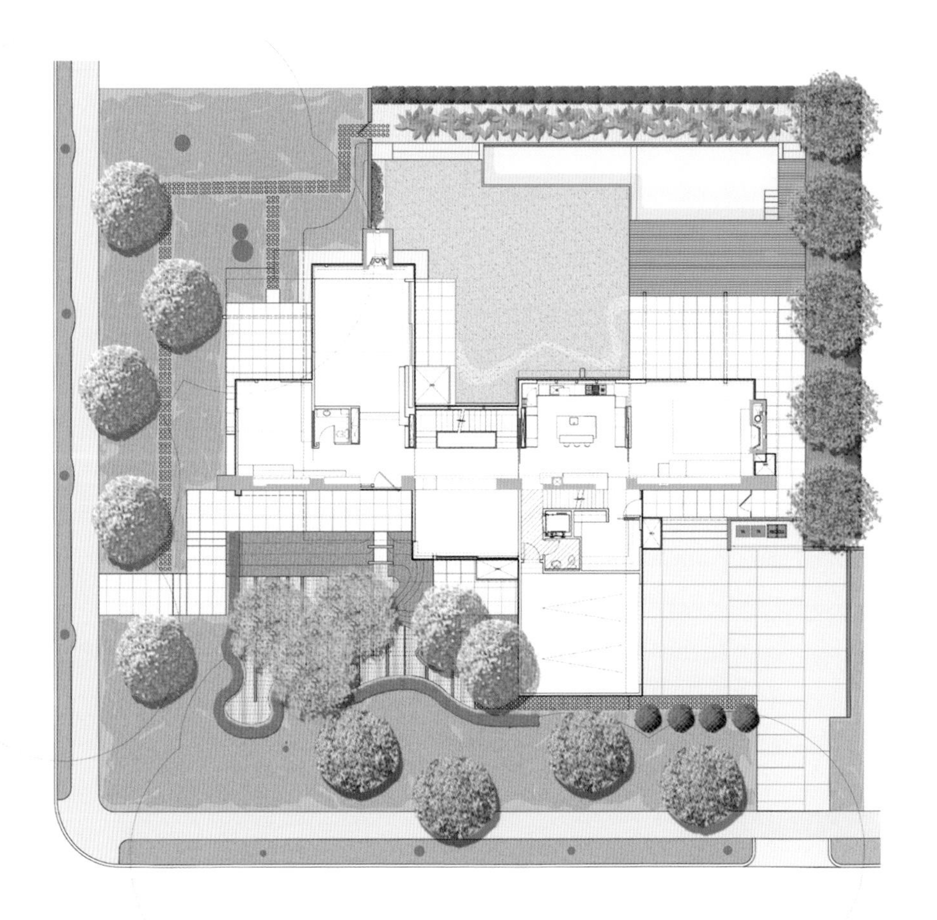

300

Peterson Residence

Design Company: Surface Design Inc.
Location: San Francisco, California U.S.A.
Photographer: Marion Brenner

Harmonious Garden Landscape with Nature

Terrain: Mountainous Region

This new garden weaves together the dual passion of the homeowners' – the desire for their children to have full range over the yard, while simultaneously displaying prominent pieces of landscape art. Through the use of grading techniques, reclaimed materials found on-site, and site-specific elements, the resulting garden is a sculpture unto itself, providing unique spaces for reflection, play, and discovery, while tying into the native, regional landscape.

Situated on a peninsula overlooking San Francisco Bay, this sculptural garden reflects the unique vision of passionate art collectors whose personal artistic impulse includes the idea of the landscape being experienced and enjoyed by their children. As a result, the garden incorporates several distinct spaces that weave together architecture, landscape, and use, while seeking to capture the site's unique views and topography.

At the site's entrance, a folding deck projects from the interior of the residence, merging the house with a landscape of magnolias, grasses, and ferns. The deck's warping planes create a threshold for the visitor, highlighting the relationship between the existing ground plane of the house and the sudden change of the surrounding topography. Beyond the deck, an existing cobbled pathway connects with a newly-planted shade garden, creating a sense of calm and intimacy around the main entrance. Opposite the main entrance path, locally-sourced river stones artfully conceal a main drainage swale, while tying into the existing slope. Past the front door, lush groves of bamboo and shade-loving trees frame the lyrical movement of a meandering path that terminates in a sculptural focal point surrounded by hillside grasses. The solidity of this hillside sculpture juxtaposes with generous sweeps of grasses, referencing the native wind-blown landscape. The approach opens up to the east lawn area, offering extensive views of San Francisco Bay and the rolling California hills to the north. Re-used granite curbing leads the way to the lower lawn where a succession of green mounds abstracts the distant hilly peninsulas that penetrate the Bay beyond. To the north of the house, a sequence of evergreen hedges and fragrant perennials create a serene viewing garden, aligning with the interior architecture. The green hedging serves as an abstracted datum highlighting the twisting movements of a kinetic sculpture by George Rickey.

Modern-style Water Garden

Skytop Road Residence

Design Company: Orange Street Studio
Location: Los Angeles, California, U.S.A.
Site Area: 3,138.6 m^2

Terrain: Sloping Fields

To create usable, private spaces in the front yard was the main design criteria in this renovation project.

16039

The challenge was to carve out these spaces from a site that sloped steeply with the street as well as towards the street. By using retaining walls, habitable and planting terraces were created from otherwise wasted land. Painted white to match the house, the retaining walls are an architectonic extension of the contemporary residence, framing planting beds and defining outdoor rooms. From street level, a flight of stairs leads one up to the front gate, comprised of an elegant wood door and translucent walls on each side. A water feature spills over ledger stone and cascades with the entry stairs.

Behind the front gate and wood wall lies a generous courtyard paved in teak wood. Here, specimen succulents surround an outdoor living room like beautiful sculptures. A fire feature acts as the centerpiece around which family and friends can gather.

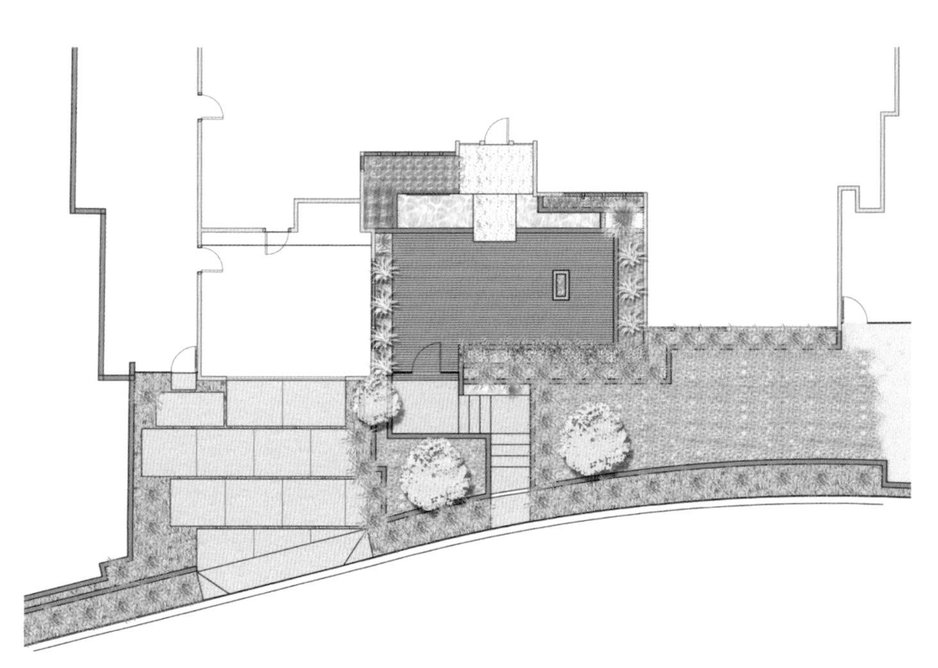

16039
16039

Adjacent to the main entry is a water feature that acts as a focal point to stop and pause before entering the house across a terrazzo bridge. By paring the plant palette to a select number of species and through the use of massing, the richness of textures and colors is highlighted. Together, the plants and warm materials create a contemporary garden that is both functional and evocative.

Seclusive Relaxed Space

Terrain: Flat Ground

This project presented the challenge of creating usable active spaces for a young couple and their children in both the front and back yards while maintaining privacy.

Napoli Drive Residence

Design Company: Orange Street Studio
Location: Los Angeles, California, USA
Site Area: 1,415.8 m^2

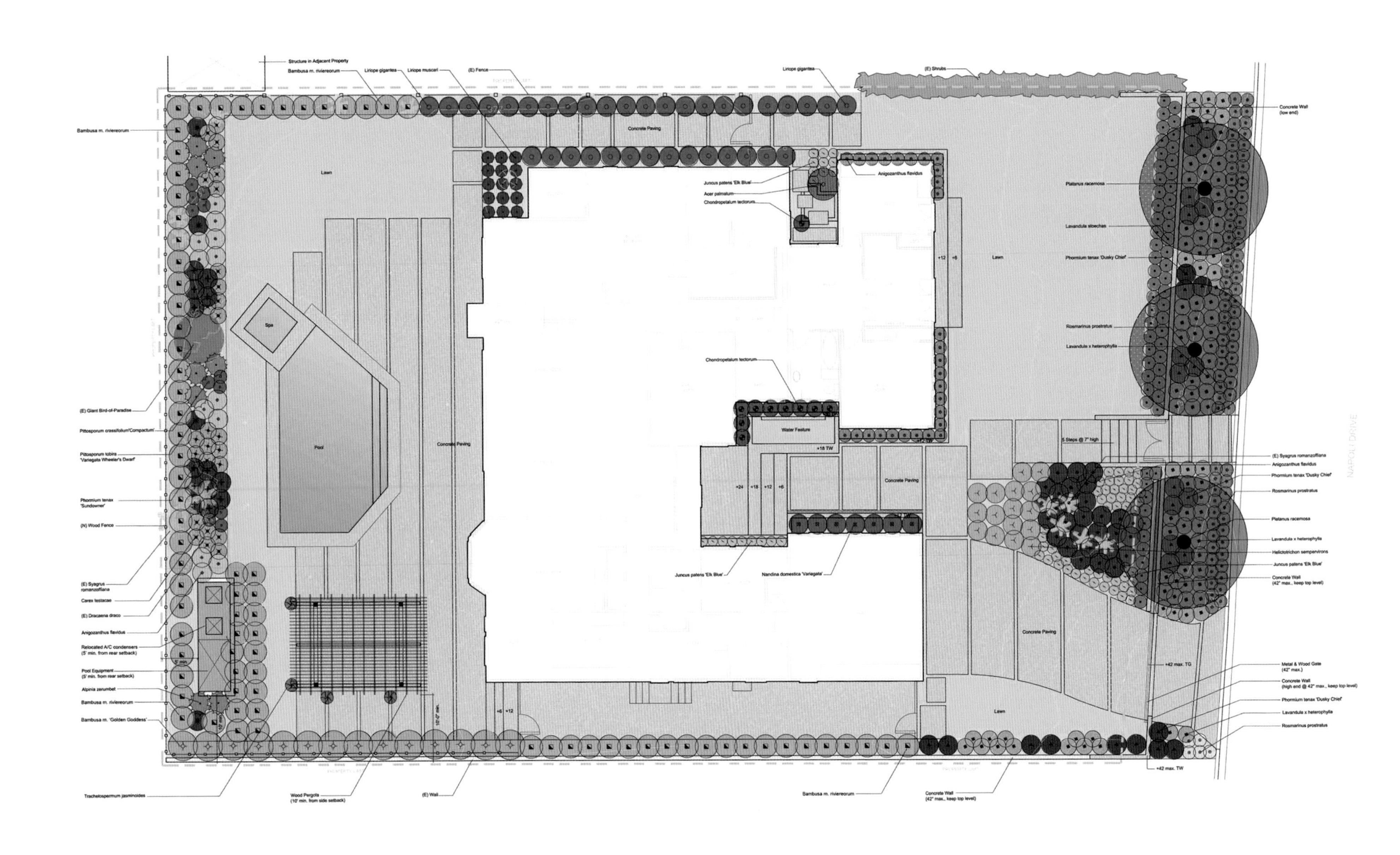

Structure in Adjacent Property
Bambusa m. riviereorum
Liriope gigantea
Liriope muscari
(E) Fence
Liriope gigantea
(E) Shrubs
Concrete Wall
(low end)
Bambusa m. riviereorum
Concrete Paving
Lawn
Juncus patens 'Elk Blue'
Acer palmatum
Chondropetalum tectorum
Anigozanthus flavidus
Platanus racemosa
Lavandula stoechas
+12
+6
Lawn
Phormium tenax 'Dusky Chief'
Spa
Rosmarinus prostratus
Lavandula x heterophylla
Chondropetalum tectorum
(E) Giant Bird-of-Paradise
Pittosporum crassifolium'Compactum'
Pittosporum tobira
'Variegata Wheeler's Dwarf'
Pool
Concrete Paving
Water Feature
+18 TW
5 Steps @ 7" high
NAPOLI DRIVE
(E) Syagrus romanzoffiana
Anigozanthus flavidus
Phormium tenax 'Dusky Chief'
+24
+18
+12
+6
Concrete Paving
Rosmarinus prostratus
Phormium tenax
'Sundowner'
(N) Wood Fence
Platanus racemosa
Lavandula x heterophylla
Helictotrichon sempervirens
Juncus patens 'Elk Blue'
Concrete Wall
(42" max., keep top level)
(E) Syagrus
romanzoffiana
Carex testacae
(E) Dracaena draco
Juncus patens 'Elk Blue'
Nandina domestica 'Variegata'
Concrete Paving
Anigozanthus flavidus
Relocated A/C condensers
(5' min. from rear setback)
5' min.
Pool Equipment
(5' min. from rear setback)
Alpinia zerumbet
Bambusa m. riviereorum
Bambusa m. 'Golden Goddess'
10' min.
+6
+12
+42 max. TG
Metal & Wood Gate
(42" max.)
Concrete Wall
(high end @ 42" max., keep top level)
Phormium tenax 'Dusky Chief'
Lavandula x heterophylla
Rosmarinus prostratus
Lawn
+42 max. TW
Trachelospermum jasminoides
Wood Pergola
(10' min. from side setback)
(E) Wall
Bambusa m. riviereorum
Concrete Wall
(42" max., keep top level)

In the front yard, a lawn play area was created by raising the topography and retaining it behind a low wall, in front of which are planted native California sycamore trees and groundcovers. This provides a safe, private play area while animating the private-public interface. Opposite a long wood bench, a water feature acts as a focal point at the main entrance to the house.

In contrast to the more public feeling of the front yard, the back yard is focused inwards, with its lush plantings and screens of bamboo on three sides of the garden. Existing palm and dragon trees are incorporated into the new landscape, which evokes a resort atmosphere of seclusion and relaxation. An existing pool and spa were renovated, and a new soaring wood arbor creates a shaded outdoor living room for family gatherings. Unifying the backyard, strips of grass break up the plane of the concrete patio. An intimate deck on the side of the house provides a serene space for outdoor showers.

Straight Style Garden

Terrain: Flat Ground

Upon completion of the house, the surrounding 500 m² property was to be designed as a garden, modern, with attractive plants and low on maintenance. It was the explicit wish of the landlord that the garden has no curves, only straight lines. A perch, with a floor-covering of dark grey flagstones and cobbles both made of granite, was created, adhering to the strict directive that everything be designing along straight lines. Hence, only use of angular cobbles and flagstones was made. The terrace adjacent to the house was built with dark granite flagstones, too. The pavement of the entry area in front of the garages consists of dark clinker bricks laid out in an edgewise fashion. The path across the lawn, connecting the entryway to the back garden runs at an acute angle, matching the water basin scheduled for completion at a later stage.

Private Suburban Garden

Designer: Alexander Oesterheld
Location: Neustadt Germany
Site Area: 500 m^2
Photographer: Ferdinand Graf von Luckner

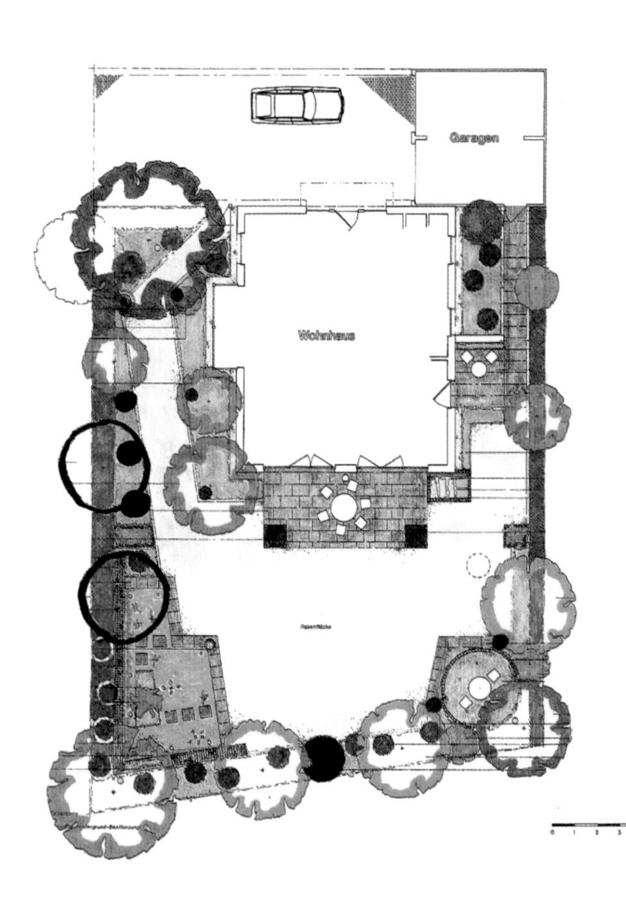
Garagen
Wohnhaus

The entire garden was surrounded by a Cherry Laurel hedge to enable proper privacy. Furthermore, to that end and to act as a windbreak, two large Austrian pines (*Pinus nigra*) were planted on the west side of the property. In addition to this, trees like Snowy mespilus (*Amelanchier lamarckii*), Snowdrop tree (*Halesia monticola*) and some fruit trees were planted serving as boundaries. Plant beds adjoining to the lawn were created to offer place to perennial plants such as Lavender (*Lavandula angustifolia*), Catmint (*Nepeta* x *faassenii*), Crane's Bill, Hostas, Ferns and different shrub-roses. A raised planting bed for culinary herbs was constructed directly next to the terrace, on the southern side of the house and an espalier apple is leaning on the house.

ORIENTAL ZEN

The representatives of the Oriental Zen are the ink painting garden of Chinese style and the succinct sketch-style courtyard of Japanese style. Affected by traditional philosophy and art of painting, the design of these gardens emphasizes the poetic feel of the landscape to achieve a perfect integration of landscapes and poetic imagery. Oriental Zen chases freshness and elegance. It stresses on accumulation of culture, pays attention to the temperament, and emphasizes the delicacy of point and plane.

Collection of Modern and Tradition

Terrain: Flat Ground

It is zen garden made in the restaurant that exists in the heartland of Old Town Kyoto. It is a traditional technique in the ZEN garden. We are expressing the river and the sea by gravel and the stone in this garden. Waterscape is expressed without using water. In addition, we expressed a modern sense by adding a sharp material such as the metals there. This garden is a space for a traditional sense and a modern sense to unite.

ZEN Garden KYOUEN

Design Company: 1moku co.
Location: Kyoto Japan
Site Area: 250 m^2
Photographer: Ryu Ishino

plan

KYOUEN　GARDEN PLAN　S=1/200　06.05.24　Imoku co.

The elements of this garden repeatedly reflect both the hand of man and nature. Water flows from the pond into the steam constructed with natural stones. Architectural wall and natural planting, geometric shape and organic shape, man cut stone arrangements and natural stone arrangements, etc. In this garden we stride to seek "absolute beauty" in change which would neither conflict or attempt to be comparative.

The garden reflects a Zen thought. This is absolute site transcending all comparative conflictions. The garden and large living room in front of the garden are in absolute unity, we have one snapshot viewed from the living room. The essence of the garden, utilizing changing grades, becomes more natural as the water flows to the lower levels, finally disappearing underground, only to resurface and flow into a formal water basin. Change of water flow expresses the essence of the Zen thought.

Cho-Rai-Tei Garden

Designer: Masuno Shunmyo
Design Company: Japan Landscape Consultants Ltd.
Location: Tokyo, Japan
Site Area: 255.5 m^2
Photographer: Tabata Minao

Natural Flowing

Terrain: Flat Ground

This residential garden located in central Tokyo is for a small family. Despite a very small space their requests for the garden were that it must have a waterfall, stream and pond that could be viewed from both the living room and tea ceremony room.

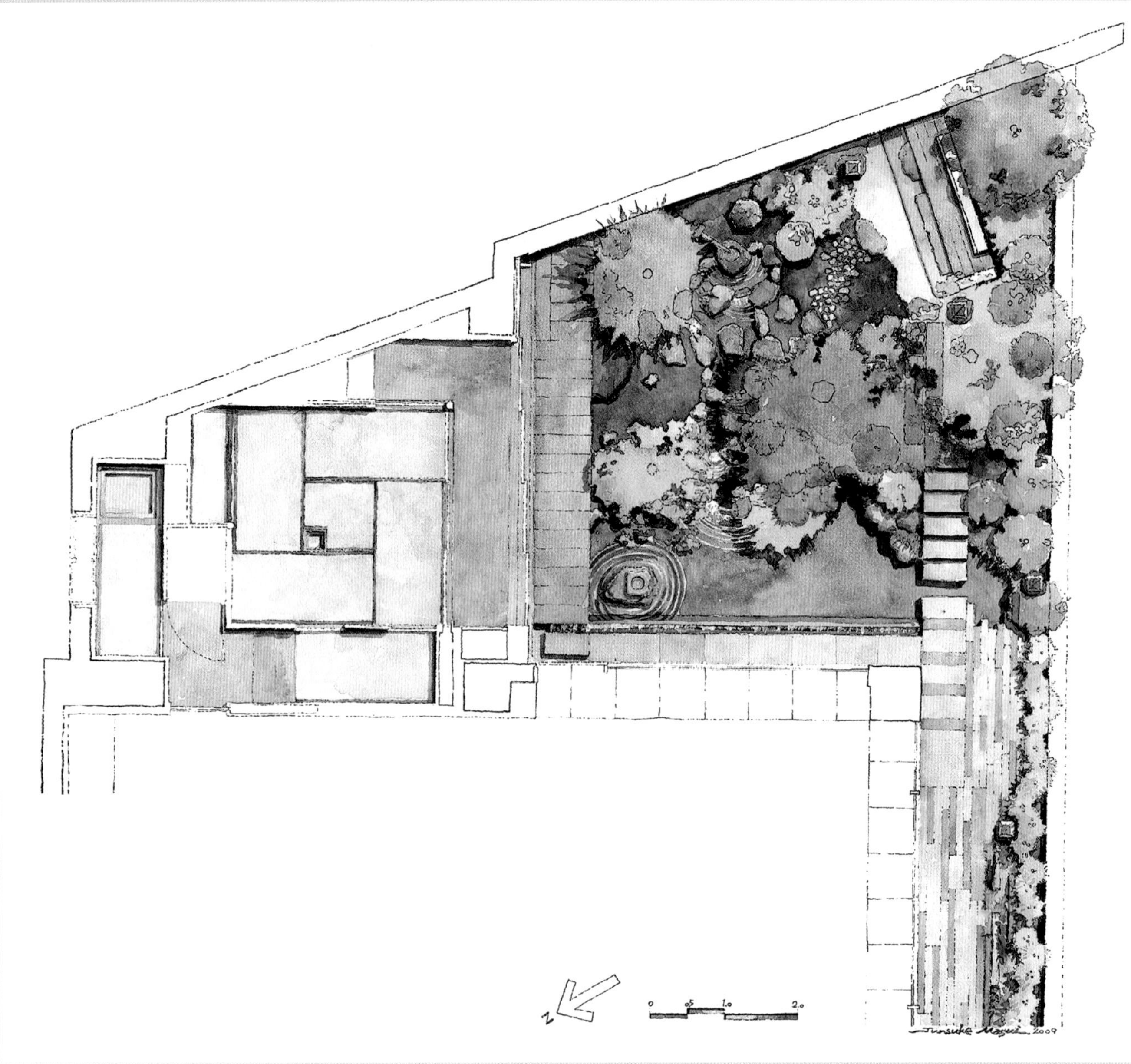

聴籟庭
Chou rai-tei

An important design element of this home was to create a relaxing, contemplative atmosphere.

As we designed the 4.5 tatami mat Kyoto style tea room as well, we were able to adjust both the room and garden to be perfectly harmonious. We created mounding behind the pond which allowed us to separate the garden functions quite naturally with plantings. A powerful black pine is the centerpiece of the planting; it allows us to feel the wind through its needles. In the far corner of the garden we designed a concrete bench house, which is in keeping with the buildings architecture.

FUSION STYLE

Fusion style shows a diversified and eclectic trend in landscape design. It integrates modern practical and traditional characteristics in the design and the layout. By utilizing many different styles, it emphasizes the general composition and visual effect of shape, color and texture.

A Fusion of Eastern and Western Styles

Terrain: Sloping Fields

Sinclair House

Design Company: Scott Brown Landscape Design
Location: Melbourne, Victoria, Australia
Site Area: 2,023.4 m^2
Photographer: Patrick Redmond

With three boys still at home, Carol and Allan's planned renovation of a 1950's brick home has benefited from the creation of a "Parent's Retreat" in every sense of the word.

Largely separated from other outdoor living areas utilized by everyone else, this courtyard has been designed to represent a microcosm of Eastern influence on our Western lives.

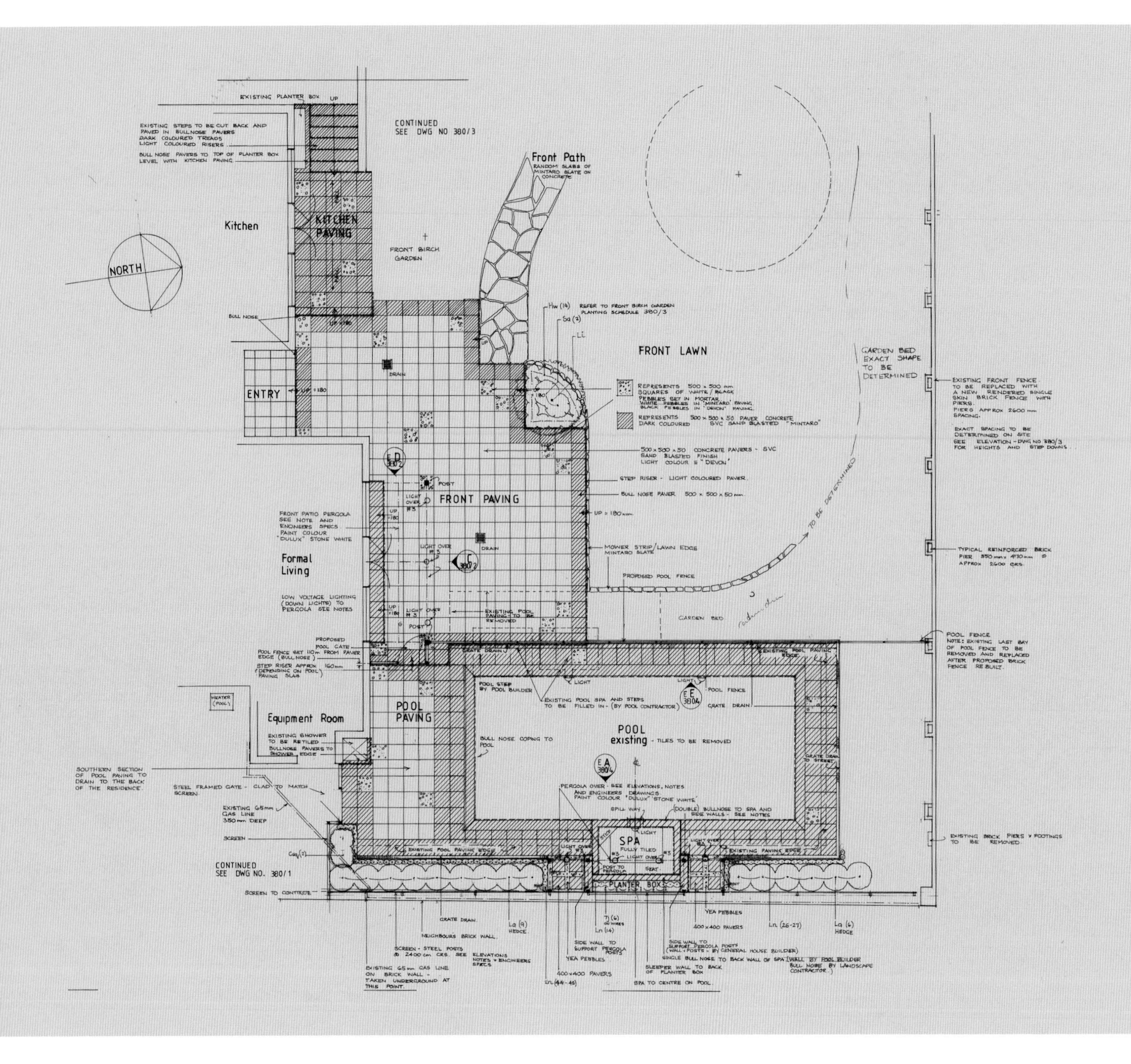

The decision to adopt an Asian-influenced design for the house itself set the scene for me create this retreat as a contemplative, restful outlook from the main bedroom, en-suite and the study.

This simple, contemplative space is made more important because it is equally enjoyed from inside the home looking out.

Given that the house floor level is significantly higher than the ground, I decided to design steps as a series of large deck landings. These landings embrace the courtyard, and contribute to the space, whilst linking it to the house itself. The result is that the outdoor space of the courtyard and the inside environment of the home seem intrinsically linked, becoming two variations of the same space. This allows the calming, restfulness of the courtyard to contribute to the ambience of the inside environment.

Structurally, the courtyard is enclosed with cement-sheet clad-walls. The walls are painted "Umbertone" grey to match the house render. This grey works well in that it allows the garden in front of this backdrop to tell the story. The plants are still differentiated and silhouetted against it, but the grey recedes and does not distract the viewer.

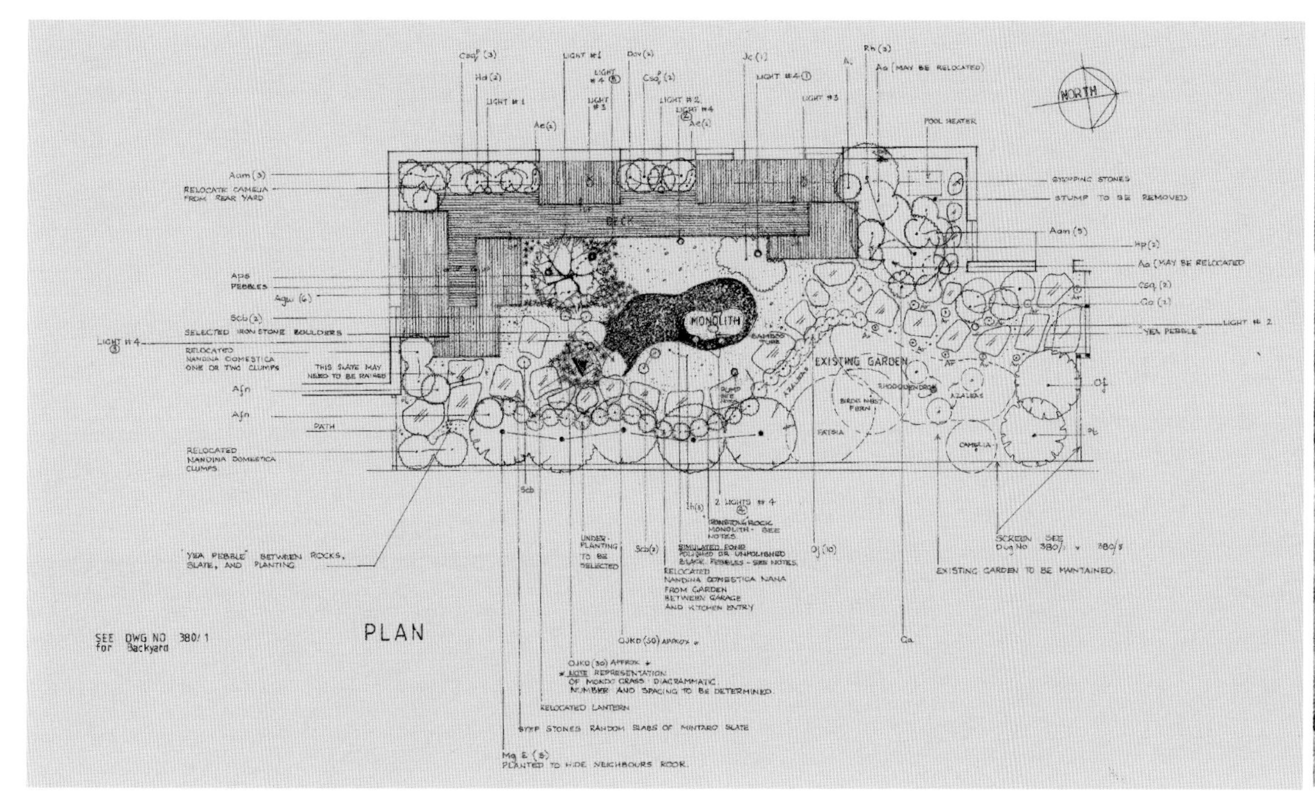

Because this space is viewed from more than one room, and from different angles, I have designed a central feature consisting of more than one focal point – each of which is visibly linked. The main central planted feature is a rare "Senkaki" maple, which is positioned at the apex of the "L" shaped deck landings. Its slightly mounded planting position is "carpeted" with massed dwarf Mondo Grass (*Ophiopogon japonicus* 'kyoto dwarf'), and it stands over the "Ikekomi-gata" lantern, as if introducing the lantern to the viewer.

At the base of these two figures, the simulated streambed (signified by grey, water-worn pebbles) meanders down slope, culminating in a small "pond" which envelopes the running-water feature. The water-feature consists of a 950 kg Ironstone boulder, with water trickling from a bamboo flute over the natural craters and cavities in the boulder, before falling through the pebbles to an underground sump, to be recycled back over the boulder. The calming sound of the gentle water movement permeates the courtyard, without dominating it or adversely affecting the tranquility that emanates from this space.

The relationship of these central elements, linked visually by the simulated streambed, is a microcosm of a much larger streambed system. This courtyard is a glimpse of a much larger pristine, natural environment... one that is captured and simplified for contemplation and respite from the stress of Carol and Allan's day-to-day lives. The use of fine, grey-white pebble surrounding the watercourse provides contrast, and relates the surrounding planted elements.

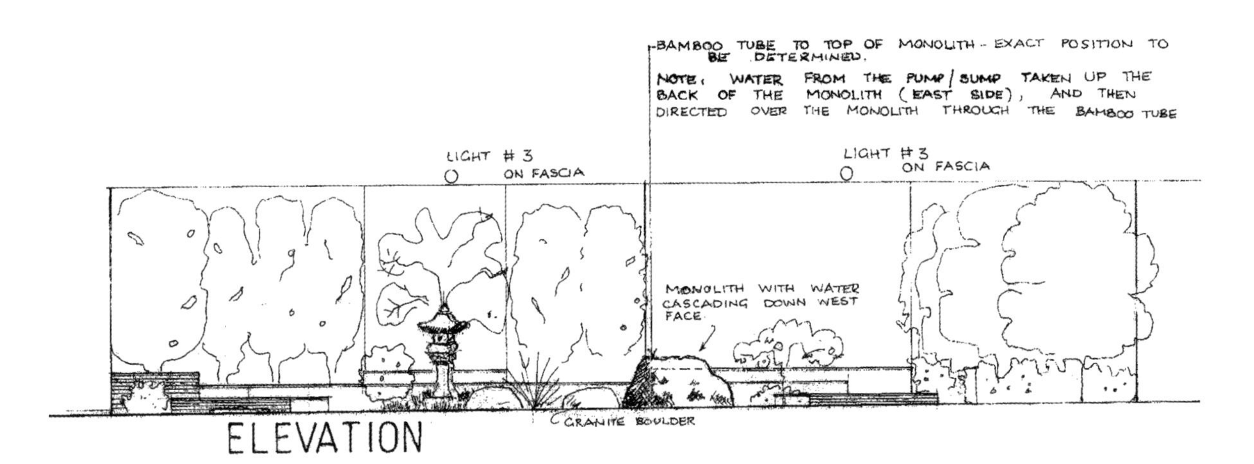

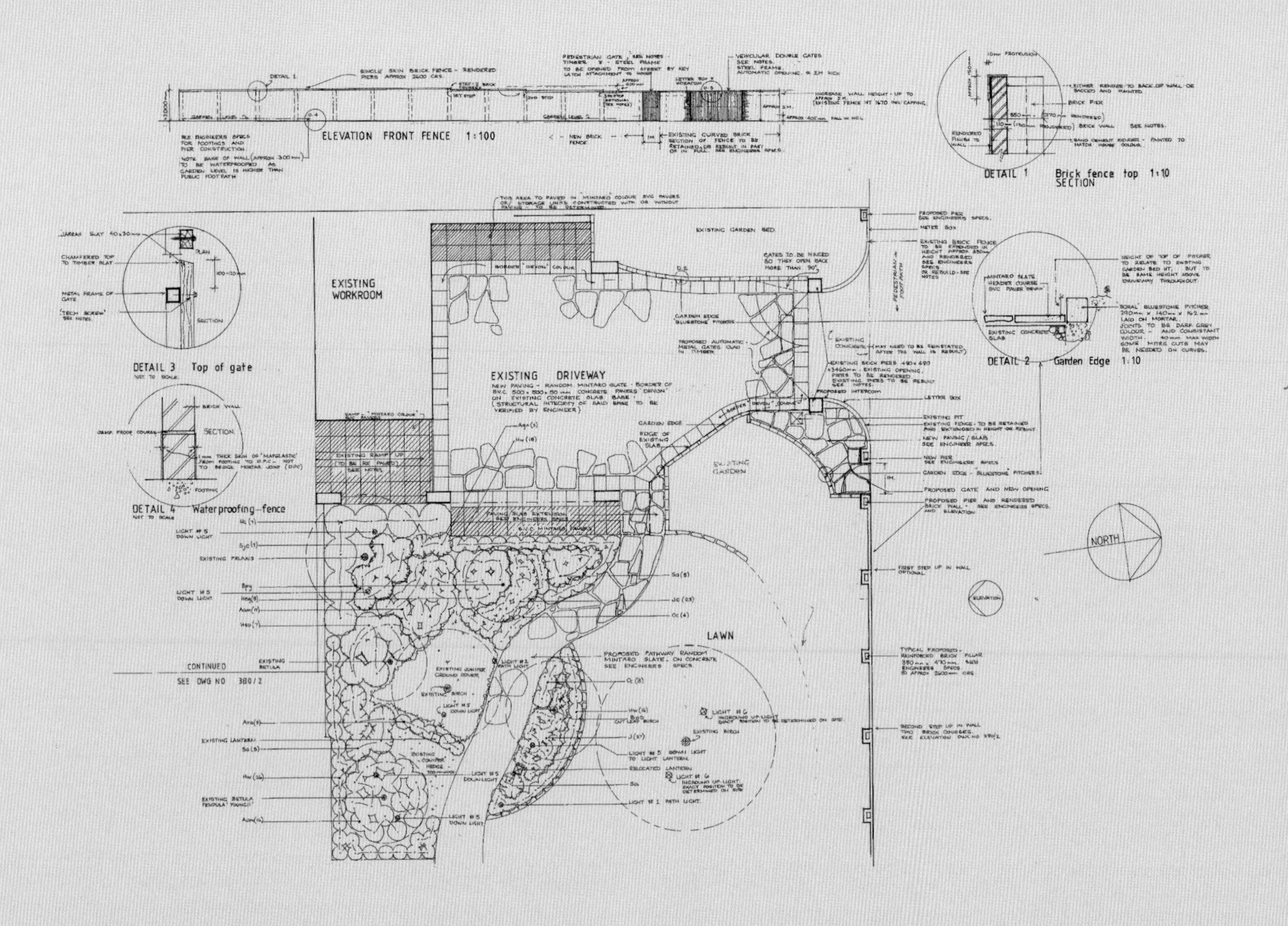

Background planting, away from the house is dominated by the repeat planting of a cultivar of evergreen Magnolia (*Magnolia grandiflora* 'Exmouth'). The bronze colour of the underneath surface of the leaves provides visual warmth and a link to the colour of the timber decks, as well as the autumnal foliage of the feature maple. The Magnolia's large leaves provide textural interest at an appropriately sensual level.

The palmate foliage of Japanese aralia (*Fatsia japonica*), combined with under-planting of rhododendrons, azaleas and Japanese sacred bamboo (*Nandina domestica*) also adds to this textural montage.

Sasanqua camelias, planted between the house walls and the decking, further soften the impact of the grey, rendered surfaces. The sasanquas have a smaller leaf size and a more compact growth habit. This differentiation in texture has played an important role in this courtyard. Large leaves and stronger foliage textures are positioned further from the house, whilst smaller, softer textures are positioned in closer proximity to the viewer. This allows the textural variation to be enjoyed without it becoming intrusive or visually distracting.

The resultant space provides a calming, restful outlook, as well as an escape from the stress of the outside world. This space is as important in its role, as any room's is within the house itself. The fact that this role is fulfilled regardless of whether one is outside the house – in the courtyard, or listening to the gentle water movement from within the sanctity of the main bedroom, make it especially valuable in Melbourne's uncooperative climate.

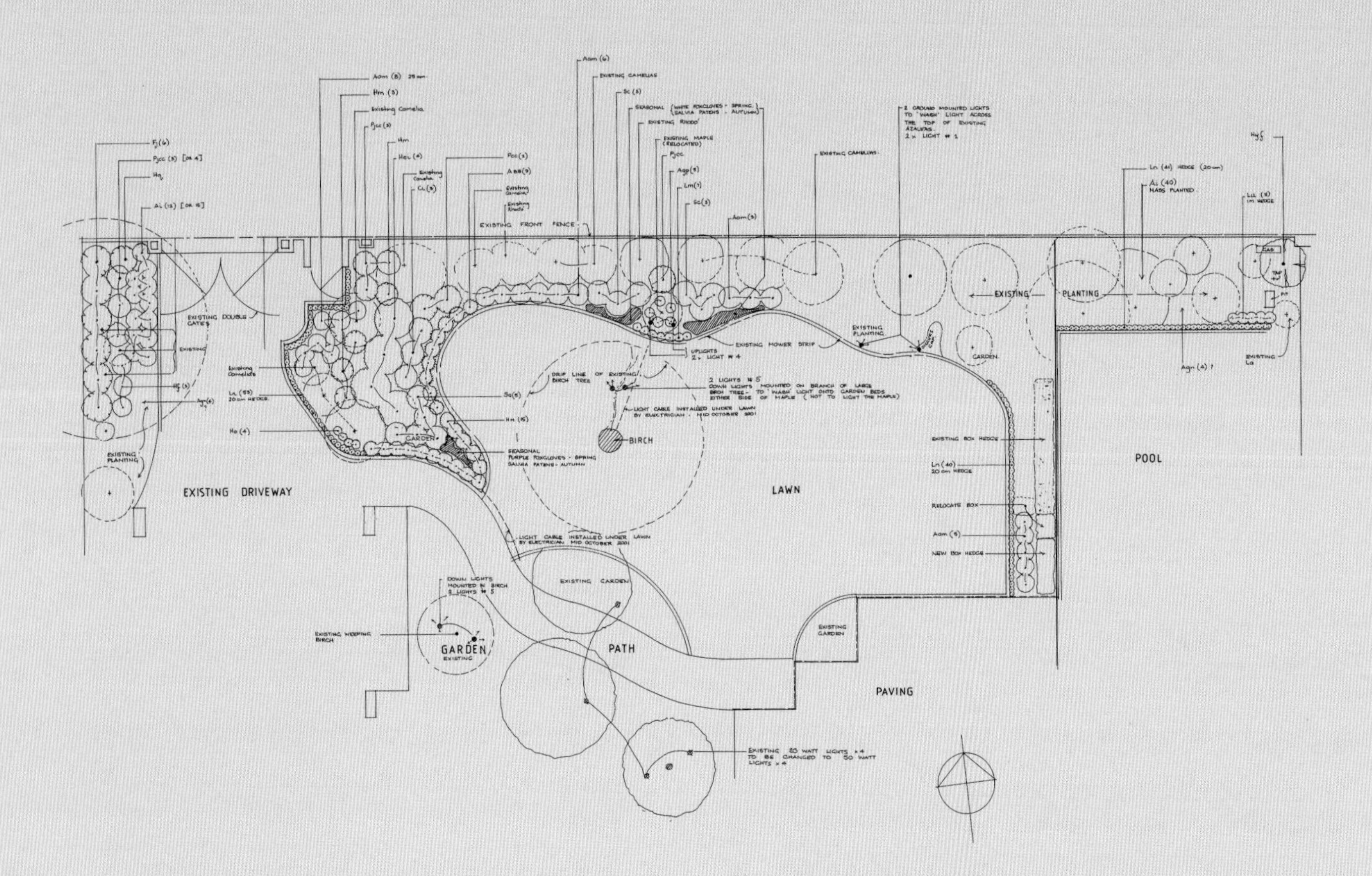

Mountainous Sea-view Villa

Terrain: Sloping Fields & Seaside

The residence had 4 distinct garden areas that required completely different treatments.

Private Residence

Design Company: Imperial Gardens Landscape Pty Ltd
Location: North of Sydney, Sydney Harbour foreshores
Site Area: 1,800 m^2

The dynamic architectural space with diagonal bridge required specific treatment for each elevation of the garden relative to the building while simultaneously creating a unified space that both reflected and enhanced the dynamic building.

The central theme was a "dry river bed" with bold rock outcrops of up to 3 tonne serpentine feature boulders. The focal point of the garden is a Japanese black pine 3 metres high, trained for 25 years with a bonsai treatment, under this is an extraordinary dynamic boulder as a cliff against white river pebbles representing the stream. To balance this a large Natsume granite water bowl was placed in a grouping of exquisite serpentine boulders and a full size japanese maple. Leading off from the japanese black pine and the bridge the stream meanders past pruned junipers , camellias , cycads and pruned Buddhist pine.

With house facing expansive ocean views there is often the need for a family to have a sheltered and protected courtyard and the small courtyard adjacent to the study was ideal for this purpose. The challenge was that this rectangular space was shapeless relative to the house. There was an opportunity to construct geometric planter beds, a semicircular stone and rendered seat and fan shaped paving that would complement the house. As the family had family ties in Fiji, it was decided to introduce a tropical selection of plants to this area. Again restrained planting was needed, yet dynamic structural plants were needed to compliment the architecture. Dracenas, rare strelitzias, gardenias and bangalow palms were used. Star jasmine and gardenias gave a fragrance to this outdoor room.

With a 50 metre drop over cliffs to the crashing waves below and the residence sitting another 20 metres higher to its curved zinc roof the site posed a considerable challenge. The client expressed his love of precise pruning, so a combination of exotic and native hedge plants were planted in the rock walls, which could then be pruned to reflect the shapes and curves of the building. Spectacular when viewed from below it was essential that the view from the house be harmonious with the stepping down to the cliffs and the sea and the distant view of the headland. Using the Japanese principle of "shakkei' or borrowed scenery a hedge was planted around the lower lawn , firstly to prevent children from going close to the cliff edge but also to be carefully shaped and pruned to harmonies with the distant headland , in effect " inviting the distant view in".

To the lower side of the house the pool area required screen planting from the neighbour. The narrow garden bed only 1 metre wide required the use of bamboos, liriopes and bamboo palms. The area was quite shaded for most of the day, yet the eastern end was subject to strong winds incompatible with bamboo. A grove of super advanced Kaizuki junipers was planted that would provide wind protection and could be sculptured through regular pruning for specific affects.

Collocated Worlds

Terrain: Mountainous Region

Built on a hill with an awe-inspiring 360° view of Beirut and the Mediterranean Sea, this garden is integrated into the natural, hilly landscape. Ranging from a modern patio with infinity pool to a variety of themed gardens, all set within an existing pine forest, this project is designed to be the perfect escape into one's own personal happiness.

From the onset, this project's most important asset was its location: a hilly area, adorned with old elegant pine trees, overlooking a bustling city arising from a calming sea that blends into a horizon. A dream location, by any standards.

The Framed Garden

Design Company: Francis Landscapes
Location: Bsalim, Lebanon
Site Area: 6,000 m^2
Photographer: Fares Jammal

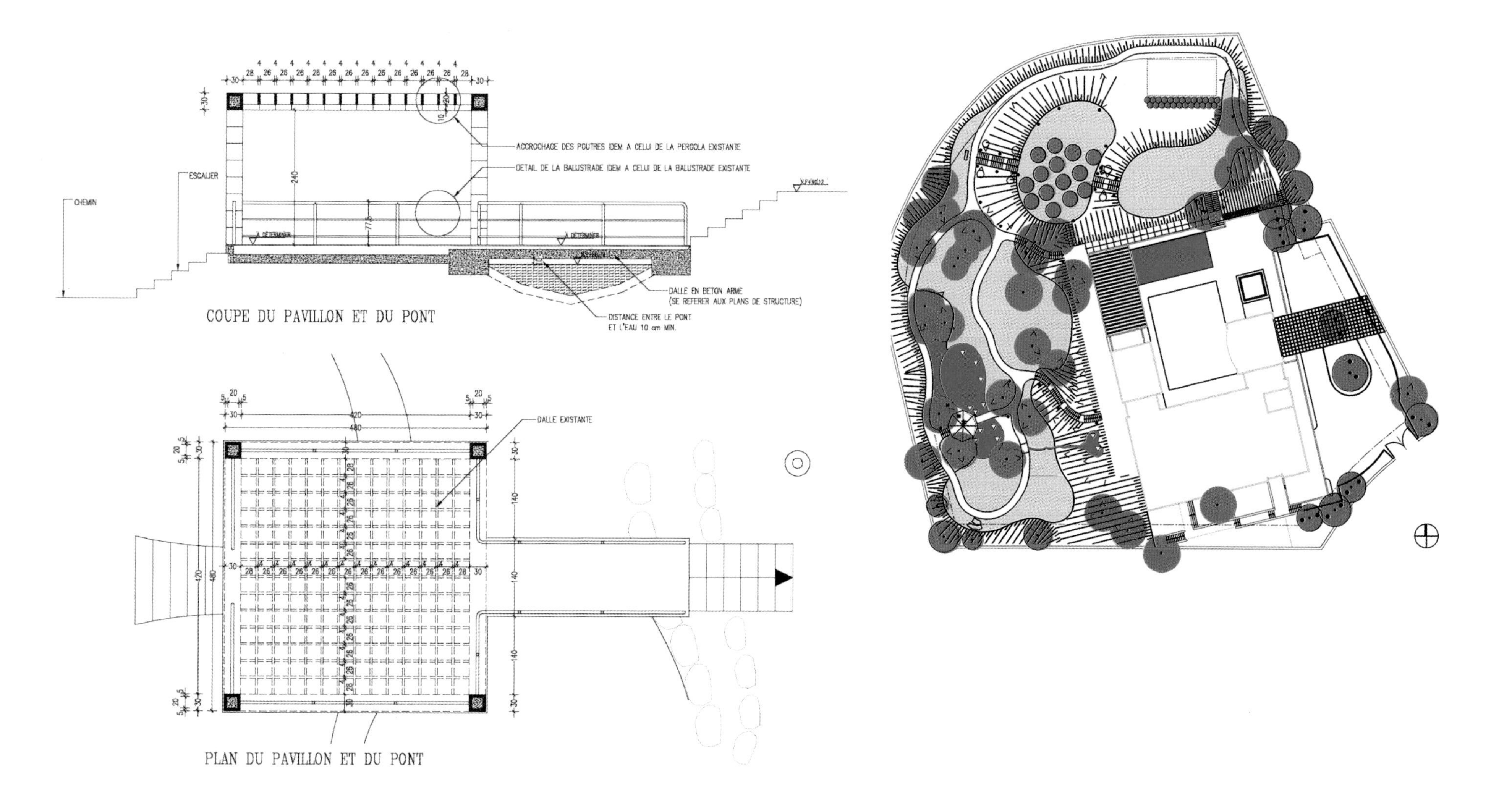

The challenge was to preserve the landscape's natural aspects, all the while integrating a modern villa within it. In order to do this, we created a patio with impressive porticos that enhance the views by framing them, transforming nature into embracing walls, vibrant with life. The elegant pine trees contrast with the structured patio walls through which they appear, like a masterpiece in a museum.

From here, the infinity pool acts like a sheet of still water, creating a dialogue by reflecting the surroundings, bestowing upon visitors an overall effect of an infinite mirror between the city below and the sky above.

From the patio, a gated walkway lures wanderers into a moment of pure serenity, exciting them emotionally and tempting them to discover the essences of the gardens unwinding before them. Below lies an exciting combination of gardens, including a Japanese garden, a bamboo garden, a rock garden, and a modern garden. This was created to be a place of contemplation and comfort, where wanderers are free to stroll through and find their own patch of personal happiness.

With one step, the wanderer leaves behind the structured world and enters a realm with lush natural forests, where all the senses are charmed, breathing with occasional clearings linked by sinuous, winding pathways. Within these collocated worlds, one's pleasures come out to be experienced through each of the senses.

Fusion Style Garden

Design Company: AguaFina Gardens International
Location: Suburbs of Detroit Michigan U.S.A.
Photographer: George Dzahristos

The Zen Garden of Chinese and Western Styles

Terrain: Flat Ground

In response to client expectations – This is the second project completed for this particular client that built this home in order to downsize from a much larger property (both in terms of overall land and square footage of the home). Their intention was to create a very low maintenance, but high impact environment with artistic details and scaled to their lifestyle. The home was actually designed around the courtyard, with most rooms overlooking this space, including the master bedroom, living room and kitchen. We had previously built a Koi pond for their former residence, and they still wanted a water feature along with visible water, but without the maintenance associated with keeping water plants and fish.